LIBRARY OF PHYSICO-CHEMICAL PROPERTY DATA

Handbook of Thermodynamic Diagrams

Volume 4

Inorganic Compounds and Elements

LIBRARY OF PHYSICO-CHEMICAL PROPERTY DATA

Handbook of Vapor Pressure
Volume 1: C_1 to C_4 Compounds (Product #5189)
Volume 2: C_5 to C_7 Compounds (Product #5190)
Volume 3: C_8 to C_{28} Compounds (Product #5191)
Volume 4: Inorganic Compounds and Elements (Product #5394)
Carl L. Yaws

Handbook of Viscosity
Volume 1: C_1 to C_4 Compounds (Product #5362)
Volume 2: C_5 to C_7 Compounds (Product #5364)
Volume 3: C_8 to C_{28} Compounds (Product #5368)
Carl L. Yaws

Handbook of Thermal Conductivity
Volume 1: C_1 to C_4 Compounds (Product #5382)
Volume 2: C_5 to C_7 Compounds (Product #5383)
Volume 3: C_8 to C_{28} Compounds (Product #5384)
Carl L. Yaws

Handbook of Thermodynamic Diagrams
Volume 1: C_1 to C_4 Compounds (Product #5857)
Volume 2: C_5 to C_7 Compounds (Product #5858)
Volume 3: C_8 to C_{28} Compounds (Product #5859)
Volume 4: Inorganic Compounds and Elements (Product #5860)
Carl L. Yaws

Each of the above series contains data for more than 1,000 organic compounds, including hydrocarbons, oxygenates, halogenates, nitrogenates, sulfur compounds, and silicon compounds. The data are presented in graphs for *vapor pressure, viscosity, thermal conductivity,* or *thermodynamics* as a function of temperature and are arranged by carbon number and chemical formula to enable the engineer to quickly determine values at the desired temperatures. Select series include wide ranges of inorganic compounds and elements.

Handbook of Transport Property Data (Product #5392)
Carl L. Yaws

Comprehensive data on viscosity, thermal conductivity, and diffusion coefficients of gases and liquids are presented in convenient tabular format.

Physical Properties of Hydrocarbons
Volume 1, Second Edition (Product #5067)
Volume 2, Third Edition (Product #5175)
Volume 3 (Product #5176)
Volume 4 (Product #5272)
R. W. Gallant and Carl L. Yaws

The four-volume series provides chemical, environmental, and safety engineers with quick and easy access to vital physical property data needed for production and process design calculations.

Thermodynamic and Physical Property Data (Product #5031)
Carl L. Yaws

Property data for 700 major hydrocarbons and organic chemicals, including oxygen, nitrogen, fluorine, chlorine, bromine, iodine, and sulfur compounds, are provided.

LIBRARY OF PHYSICO-CHEMICAL PROPERTY DATA

Handbook of Thermodynamic Diagrams

Volume 4

Inorganic Compounds and Elements

Carl L. Yaws

Gulf Publishing Company

Handbook of Thermodynamic Diagrams, Volume 4

Gulf Publishing Company
Book Division
P.O. Box 2608 ☐ Houston, Texas 77252-2608

Printed and bound in the United Kingdom
Transfered to Digital Printing, 2011
Printed on Acid-Free Paper (∞)

Library of Congress Cataloging-in-Publication Data

Yaws, Carl L.
    Handbook of thermodynamic diagrams : volume and enthalpy diagrams for major organic chemicals and hydrocarbons / Carl L. Yaws.
        p.   cm.—(Library of physico-chemical property data)
    Includes bibliographical references.
    Contents: v. 1. C_1 to C_4 compounds — v. 2. C_5 to C_7 compounds — v. 3. C_8 to C_{28} compounds — v. 4. Inorganic compounds and elements.
    ISBN 0-88415-857-8 (v. 1 : alk. paper). — ISBN 0-88415-858-6 (v. 2 : alk. paper). — ISBN 0-88415-859-4 (v. 3 : alk. paper). — **ISBN 0-88415-860-8** (v. 4 : alk. paper)
    1. Thermodynamics—Tables.   2. Hydrocarbons—Tables.   3. Organic compounds—Tables.   I. Title. II. Series.
QD504.Y36   1996
660'.2969'0223—dc20   96-36328
                                                          CIP

CONTENTS

CONTRIBUTORS

Mei Han

Graduate student, Chemical Engineering Department, Lamar University, Beaumont, Texas 77710, U.S.A.

Sachin D. Sheth

Graduate Student, Chemical Engineering Department, Lamar University, Beaumont, Texas 77710, U.S.A.

Carl L. Yaws

Professor, Chemical Engineering Department, Lamar University, Beaumont, Texas 77710, U.S.A.

ACKNOWLEDGMENTS

Many colleagues and students have made contributions and helpful comments over the years. The author is grateful to each: Jack R. Hopper, Joe W. Miller, Jr., C. S. Fang, K. Y. Li, Keith C. Hansen, Daniel H. Chen, Fu-Ming Tsuo, Jeng-Shia Cheng, San-Min Chou, J. W. Via, III, P. Y. Chiang, H. C. Yang, Xiang Pan, Xiaoyan Lin, Duane G. Piper, Jr., Li Bu, Sachin D. Nijhawan, Sachin D. Sheth, and Mei Han.

The author wishes to acknowledge special appreciation to his wife (Annette) and family (Kent, Michele, Chelsea, and Brandon; Lindsay and Rebecca; and Matthew and Sarah).

The author wishes to acknowledge that the Gulf Coast Hazardous Substance Research Center provided partial support to this work.

DISCLAIMER

This handbook presents a variety of thermodynamic and physical property data. It is incumbent upon the user to exercise judgment in the use of the data. The author and publisher do not provide any guarantee, express or implied, with regard to the general or specific applicability of the data, the range of errors that may be associated with any of the data, or the appropriateness of using any of the data in any subsequent calculation, design, or decision process. The author and publisher accept no responsibility for damages, if any, suffered by any reader or user of this handbook as a result of decisions made or actions taken on information contained herein.

PREFACE

Thermodynamic property data are important in many engineering applications in the chemical processing and petroleum refining industries. The objective of this book is to provide the engineer with such data. The data are presented in thermodynamic diagrams (graphs) covering a wide range of pressures and temperatures to enable the engineer to quickly determine values at points of interest. The contents of the book are arranged in the following order: graphs, references, and appendixes.

The graphs are arranged by carbon number and chemical formula to provide ease of use. English units are used for the property values. For those involved in SI and metric usage, each graph displays a conversion factor to provide the SI and metric units.

The graphs provide wide coverage for volume and enthalpy as a function of temperature and pressure, including the following:

- two-phase region for saturated liquid and vapor
- superheated gas region for gases above saturation temperature
- subcooled liquid region for liquids below saturation temperature
- supercritical region for temperatures and pressures above critical point

The graphs for enthalpy also contain lines of constant entropy to permit engineering usage for 2nd law problems such as adiabatic expansion and compression of fluids.

The coverage encompasses a wide range of compounds (total = 343). The coverage of inorganics is comprehensive: carbon oxides, such as carbon monoxide and carbon dioxide; nitrogen oxides, such as nitric oxide and nitrous oxide; sulfur oxides, such as sulfur dioxide and sulfur trioxide; hydrogen oxides, such as water and hydrogen peroxide; ammonias, such as ammonia and ammonium hydroxide; hydrogen halides, such as hydrogen chloride and hydrogen fluoride; sulfur acids, such as sulfuric acid and hydrogen sulfide; hydroxides, such as sodium hydroxide and potassium hydroxide; silicon halides, such as trichlorosilane and silicon tetrachloride; ureas, such as urea and thiourea; cyanides, such as hydrogen cyanide and cyanogen chloride; hydrides, such as silane and diborane; sodium derivatives, such as sodium chloride and sodium fluoride; aluminum derivatives, such as aluminum borohydride and aluminum fluoride; and many other compound types. Many elements (total = 82) are covered: hydrogen, nitrogen, oxygen, helium, argon, neon, chlorine, bromine, iodine, fluorine, sulfur, phosphorous, aluminum, lead, tin, mercury, sodium, magnesium, silicon, antimony, boron, iron, chromium, cobalt, titanium, tantalum, silver, gold, platinum, radon, uranium, and many others.

For most compounds, the range of coverage for pressure is from 10 to 10,000 psia. Very limited experimental data are available at pressures above 1,000 to 2,000 psia. Thus, values at the higher pressures should be considered rough approximations. Values at lower pressures are more accurate.

The graphs are based on the Peng-Robinson equation of state (1) as improved by Stryjek and Vera (2, 3). The equations for thermodynamic properties using the Peng-Robinson equation of state are given in the appendix for volume, compressibility factor, fugacity coefficient, residual enthalpy, and residual entropy. Critical constants and ideal gas heat capacities for use in the equations are from the data compilations of DIPPR (8) and Yaws (28, 29, 30).

The literature has been carefully searched in construction of the graphs. References for sources used in preparing the work are given in the section following the graphs near the end of the book.

For the graphs, some of the compounds may undergo thermal decomposition (reaction) at the higher temperatures. For such cases of thermal decomposition, the graphs are useful for ascertaining property values of the pure compound which is contained in the reaction mixture. Chemistry handbooks and DIPPR (8) notes may be used for specifics regarding thermal decomposition.

A list of compounds is given near the end of the book to aid the user in quickly locating compounds of interest from knowledge of the chemical formula or name.

An executable computer program, complete with data files, is available for calculation of thermodynamic properties. For information on the program, contact Carl L. Yaws, Ph.D., P. O. Box 10053, Beaumont, Texas 77710, phone/fax (409) 880-8787.

LIBRARY OF PHYSICO-CHEMICAL PROPERTY DATA

Handbook of Thermodynamic Diagrams

Volume 4

Inorganic Compounds and Elements

1. Molecular Weight, lb/mol.......... 107.868

2. Freezing Point, F........................ 1761.5

3. Boiling Point, F........................ 4013.3

4. Density @ 20 C, g/cm^3............ 10.5

5. Density @ 68 F, lb/ft^3.............. 655.49

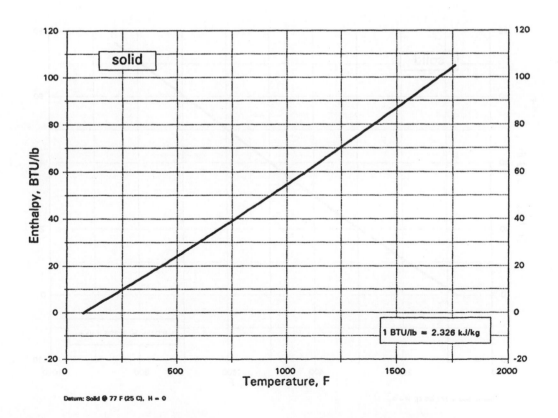

solid

1 BTU/lb = 2.326 kJ/kg

Datum: Solid @ 77 F (25 C), H = 0

1

1. Molecular Weight, lb/mol.......... 143.321

2. Freezing Point, F........................ 851

3. Boiling Point, F........................... 2847.2

4. Density @ 20 C, g/cm^3............ 5.56

5. Density @ 68 F, lb/ft^3.............. 347.1

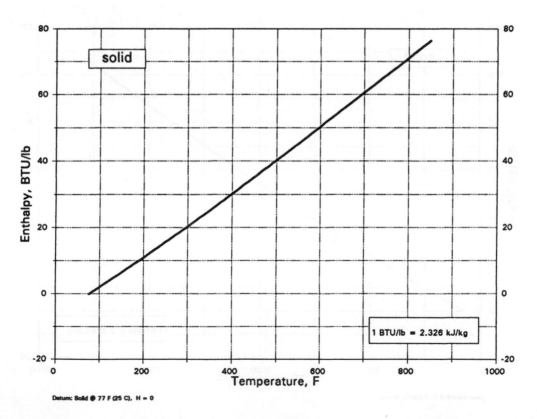

solid

1 BTU/lb = 2.326 kJ/kg

Enthalpy, BTU/lb

Temperature, F

Datum: Solid @ 77 F (25 C), H = 0

Ag	SILVER

1. Molecular Weight, lb/mol.......... 107.868

2. Freezing Point, F....................... 1761.5

3. Boiling Point, F.......................... 4013.3

4. Density @ 20 C, g/cm^3............. 10.5

5. Density @ 68 F, lb/ft^3............... 655.49

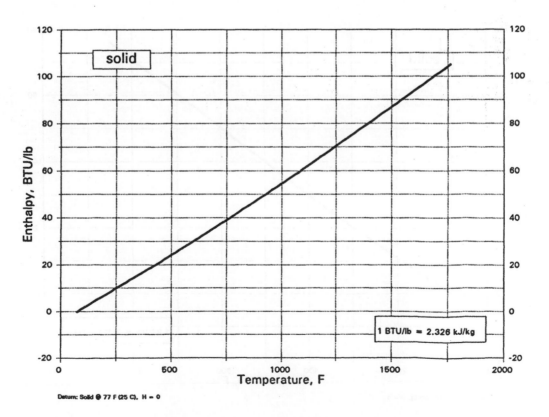

solid

1 BTU/lb = 2.326 kJ/kg

Datum: Solid @ 77 F (25 C), H = 0

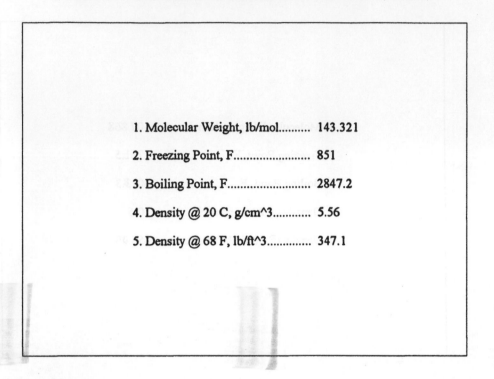

1. Molecular Weight, lb/mol.......... 143.321

2. Freezing Point, F........................ 851

3. Boiling Point, F.......................... 2847.2

4. Density @ 20 C, g/cm^3............. 5.56

5. Density @ 68 F, lb/ft^3.............. 347.1

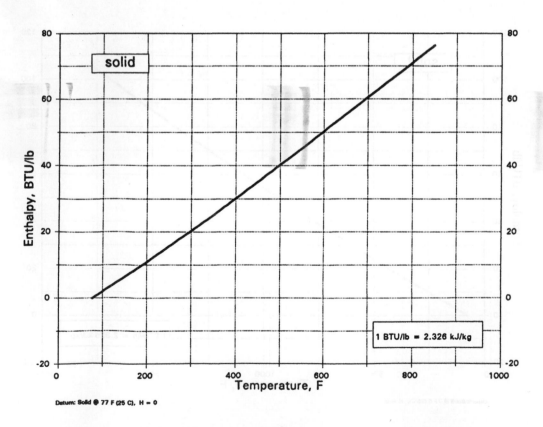

solid

1 BTU/lb = 2.326 kJ/kg

Enthalpy, BTU/lb

Temperature, F

Datum: Solid @ 77 F (25 C), H = 0

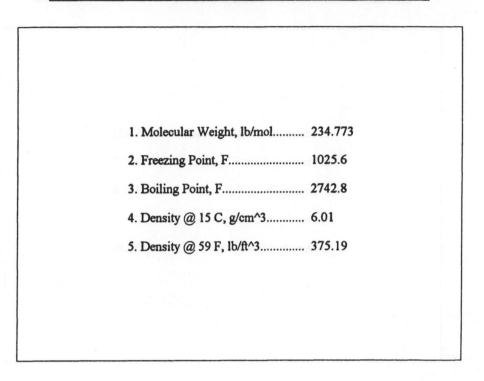

1. Molecular Weight, lb/mol.......... 234.773

2. Freezing Point, F....................... 1025.6

3. Boiling Point, F......................... 2742.8

4. Density @ 15 C, g/cm^3............ 6.01

5. Density @ 59 F, lb/ft^3.............. 375.19

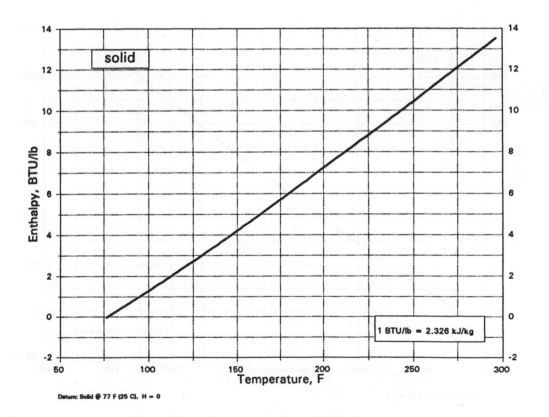

Datum: Solid @ 77 F (25 C), H = 0

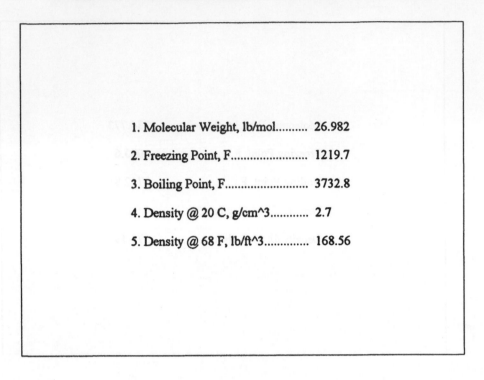

1. Molecular Weight, lb/mol.......... 26.982

2. Freezing Point, F...................... 1219.7

3. Boiling Point, F......................... 3732.8

4. Density @ 20 C, g/cm^3............ 2.7

5. Density @ 68 F, lb/ft^3.............. 168.56

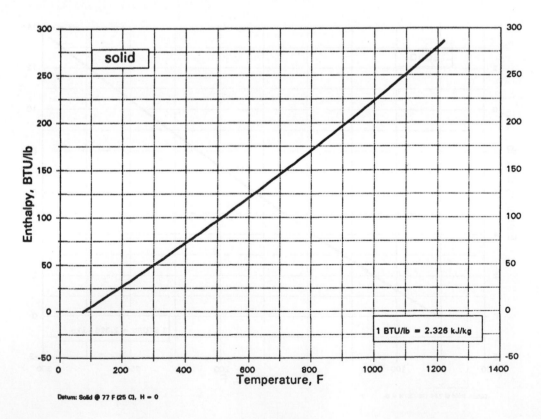

solid

1 BTU/lb = 2.326 kJ/kg

Datum: Solid @ 77 F (25 C), H = 0

1. Molecular Weight, lb/mol.......... 71.51

2. Freezing Point, F........................ -83.2

3. Boiling Point, F......................... 114.6

4. Density @ C, g/cm^3............. ---

5. Density @ F, lb/ft^3.............. ---

1. Molecular Weight, lb/mol.......... 71.51

2. Freezing Point, F........................ -83.2

3. Boiling Point, F......................... 114.6

4. Density @ C, g/cm^3............. ---

5. Density @ F, lb/ft^3.............. ---

Heat capacity data are not available.

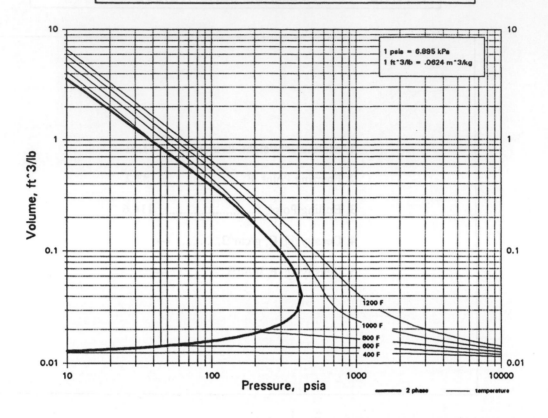

1 psia = 6.895 kPa
1 ft^3/lb = .0624 m^3/kg

1200 F
1000 F
800 F
600 F
400 F

2 phase   temperature

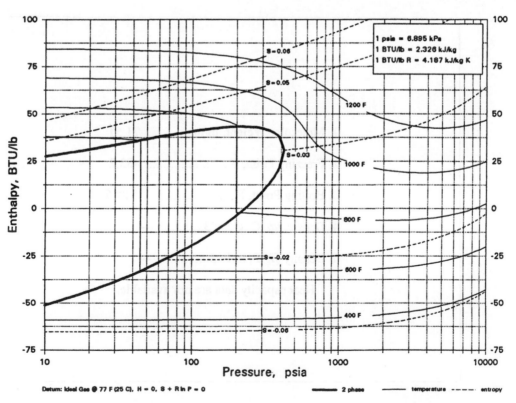

1 psia = 6.895 kPa
1 BTU/lb = 2.326 kJ/kg
1 BTU/lb R = 4.187 kJ/kg K

S = 0.06
S = 0.05
S = 0.03
S = -0.02
S = -0.06

1200 F
1000 F
800 F
600 F
400 F

Datum: Ideal Gas @ 77 F (25 C), H = 0, S + R ln P = 0

2 phase   temperature   ----- entropy

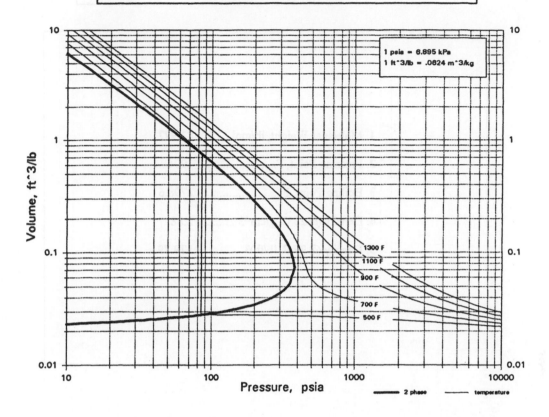

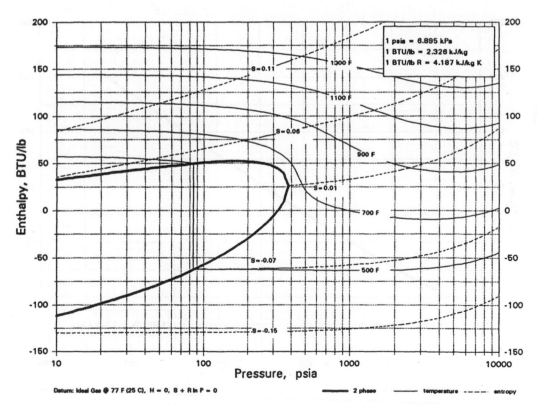

1. Molecular Weight, lb/mol.......... 83.977

2. Freezing Point, F...................... 1904

3. Boiling Point, F......................... 2798.6

4. Density @ 25 C, g/cm^3............. 2.88

5. Density @ 77 F, lb/ft^3.............. 179.79

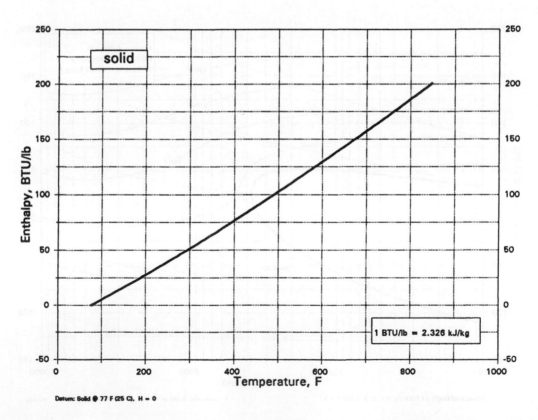

Datum: Solid @ 77 F (25 C), H = 0

1. Molecular Weight, lb/mol.......... 407.695

2. Freezing Point, F........................ 375.8

3. Boiling Point, F......................... 725.9

4. Density @ 25 C, g/cm^3............. 3.98

5. Density @ 77 F, lb/ft^3.............. 248.46

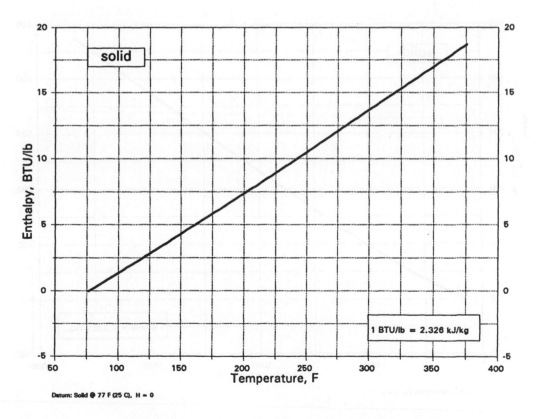

Datum: Solid @ 77 F (25 C), H = 0

1. Molecular Weight, lb/mol.......... 101.961

2. Freezing Point, F....................... 3725.3

3. Boiling Point, F......................... 5396

4. Density @ 25 C, g/cm^3............ 3.97

5. Density @ 77 F, lb/ft^3.............. 247.84

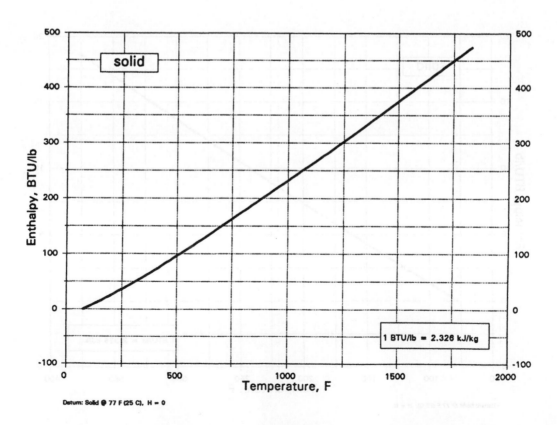

solid

1 BTU/lb = 2.326 kJ/kg

Datum: Solid @ 77 F (25 C), H = 0

1. Molecular Weight, lb/mol.......... 342.154

2. Freezing Point, F...................... 1418.1

3. Boiling Point, F......................... ---

4. Density @ 20 C, g/cm^3............ 2.71

5. Density @ 68 F, lb/ft^3.............. 169.18

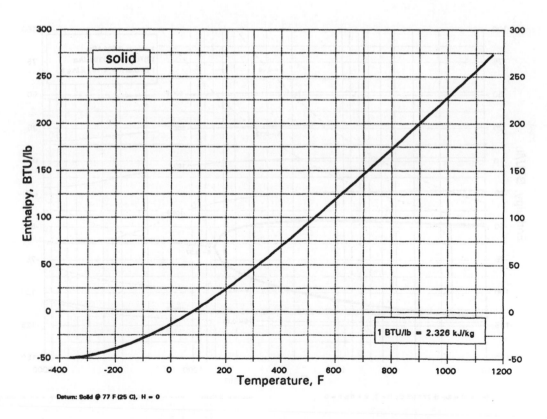

Datum: Solid @ 77 F (25 C), H = 0

11

Ar   ARGON

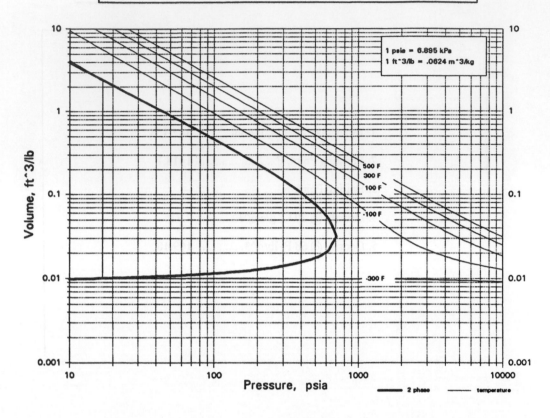

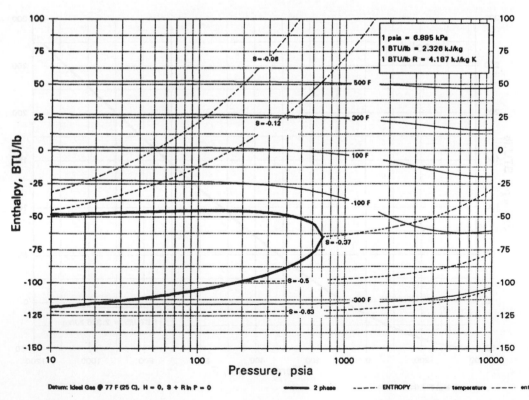

1. Molecular Weight, lb/mol.......... 74.922

2. Freezing Point, F....................... 1502.6

3. Boiling Point, F........................ 1133.3

4. Density @ 14 C, g/cm^3............ 5.73

5. Density @ 57 F, lb/ft^3.............. 357.71

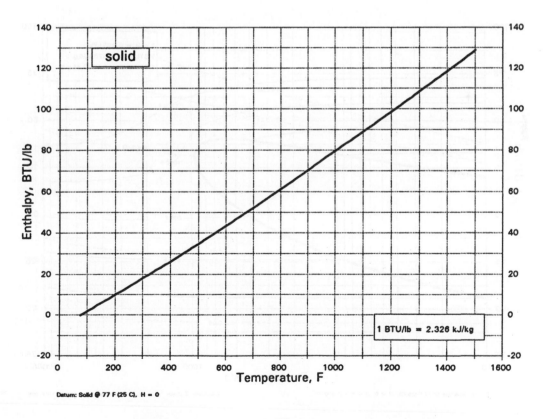

solid

1 BTU/lb = 2.326 kJ/kg

Datum: Solid @ 77 F (25 C), H = 0

13

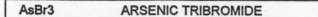

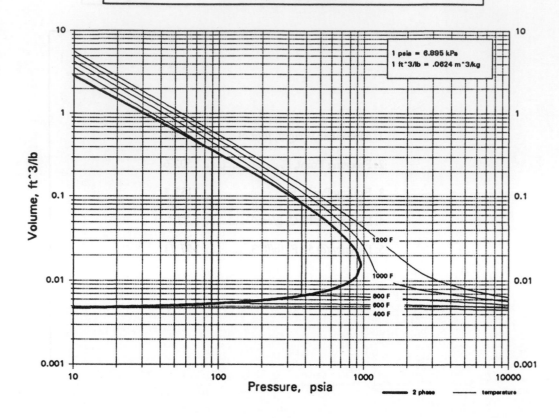

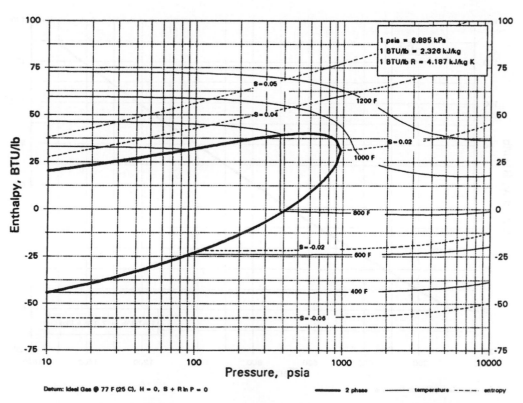

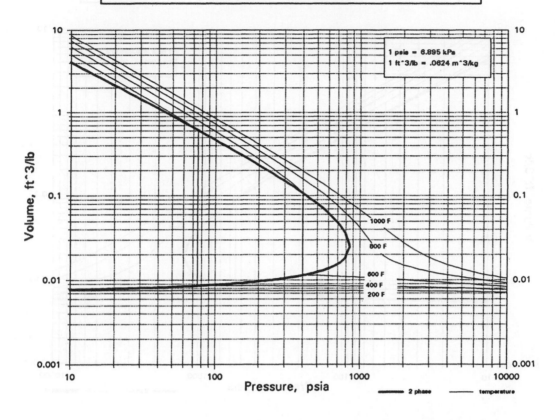

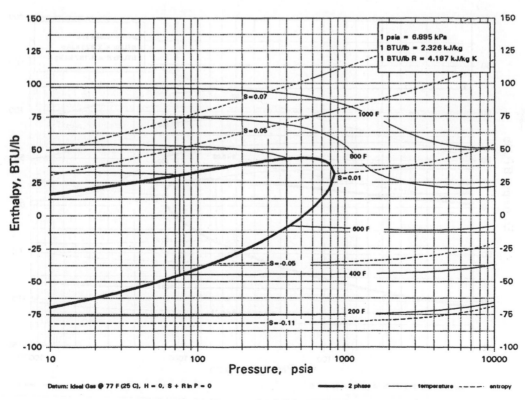

Datum: Ideal Gas @ 77 F (25 C), H = 0, S + R ln P = 0

15

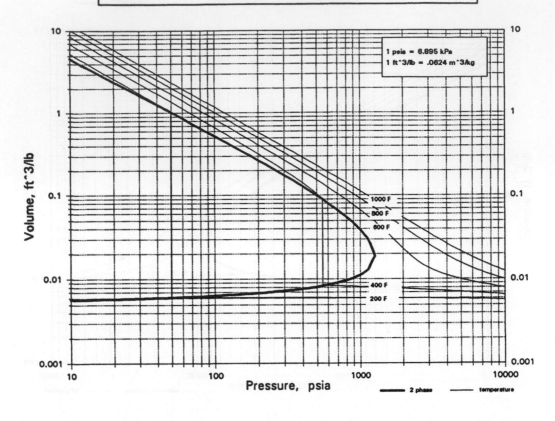

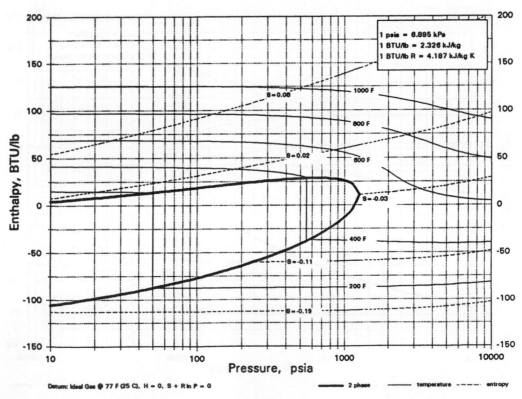

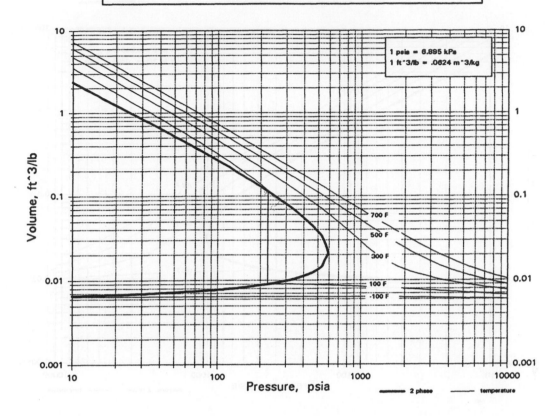

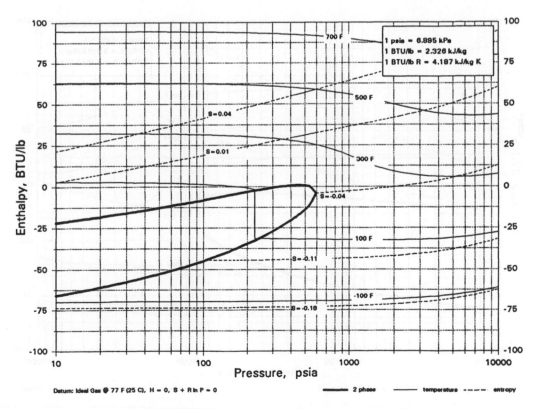

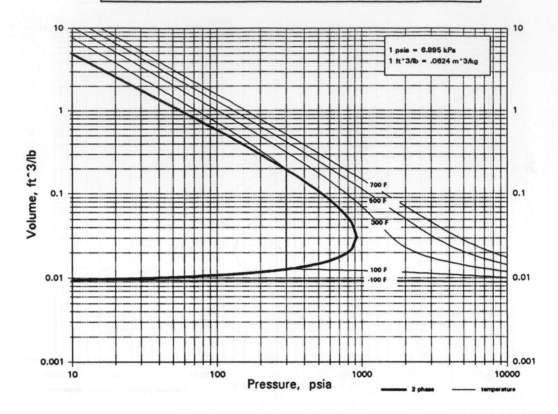

1 psia = 6.895 kPa
1 ft^3/lb = .0624 m^3/kg

2 phase   temperature

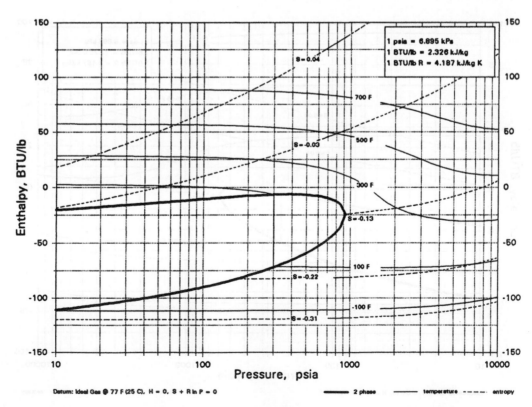

1 psia = 6.895 kPa
1 BTU/lb = 2.326 kJ/kg
1 BTU/lb R = 4.187 kJ/kg K

Datum: Ideal Gas @ 77 F (25 C), H = 0, S + R ln P = 0

2 phase   temperature   entropy

18

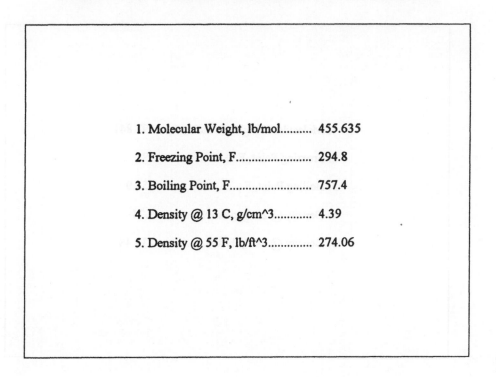

1. Molecular Weight, lb/mol.......... 455.635

2. Freezing Point, F....................... 294.8

3. Boiling Point, F......................... 757.4

4. Density @ 13 C, g/cm^3............. 4.39

5. Density @ 55 F, lb/ft^3.............. 274.06

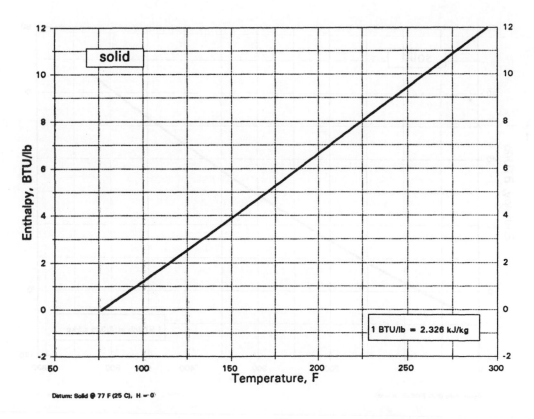

1 BTU/lb = 2.326 kJ/kg

Datum: Solid @ 77 F (25 C), H = 0

1. Molecular Weight, lb/mol.......... 197.841

2. Freezing Point, F....................... 595

3. Boiling Point, F.......................... 854.9

4. Density @ 20 C, g/cm^3............ 3.74

5. Density @ 68 F, lb/ft^3.............. 233.48

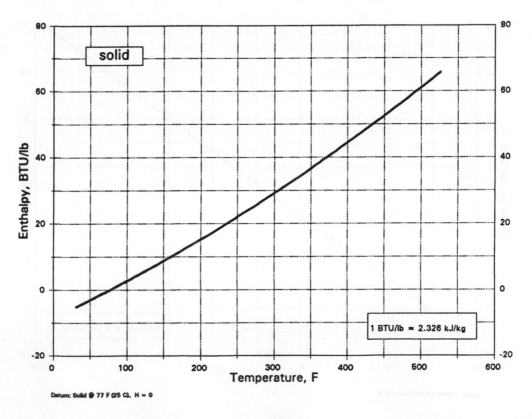

solid

1 BTU/lb = 2.326 kJ/kg

Enthalpy, BTU/lb

Temperature, F

Datum: Solid @ 77 F (25 C), H = 0

20

1. Molecular Weight, lb/mol.......... 210

2. Freezing Point, F........................ 575.6

3. Boiling Point, F.......................... 632.9

4. Density @ C, g/cm^3............. ---

5. Density @ F, lb/ft^3............. ---

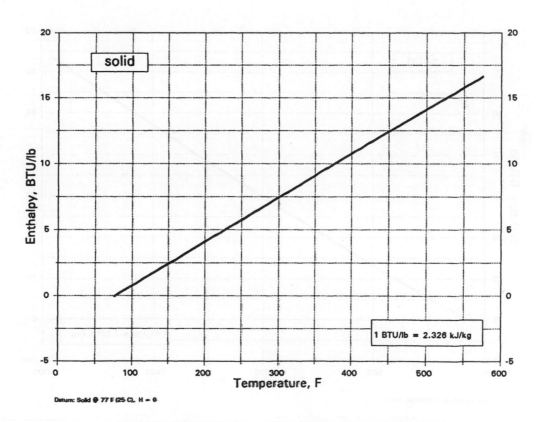

solid

Enthalpy, BTU/lb

Temperature, F

1 BTU/lb = 2.326 kJ/kg

Datum: Solid @ 77 F (25 C), H = 0

21

1. Molecular Weight, lb/mol.......... 196.967

2. Freezing Point, F...................... 1947.5

3. Boiling Point, F........................ 5156.3

4. Density @ 20 C, g/cm^3............ 19.31

5. Density @ 68 F, lb/ft^3.............. 1205.48

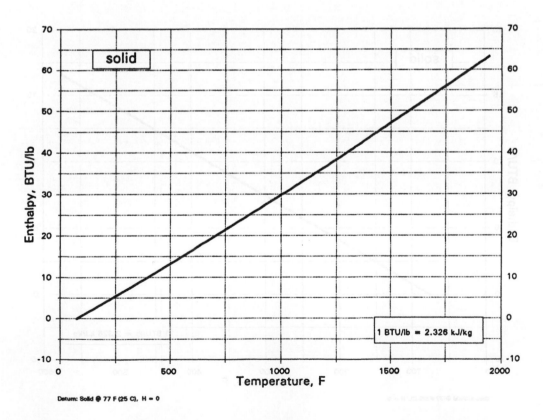

solid

1 BTU/lb = 2.326 kJ/kg

Datum: Solid @ 77 F (25 C), H = 0

1. Molecular Weight, lb/mol.......... 10.811

2. Freezing Point, F....................... 3767

3. Boiling Point, F.......................... 6979.7

4. Density @ 20 C, g/cm^3............ 2.34

5. Density @ 68 F, lb/ft^3.............. 146.08

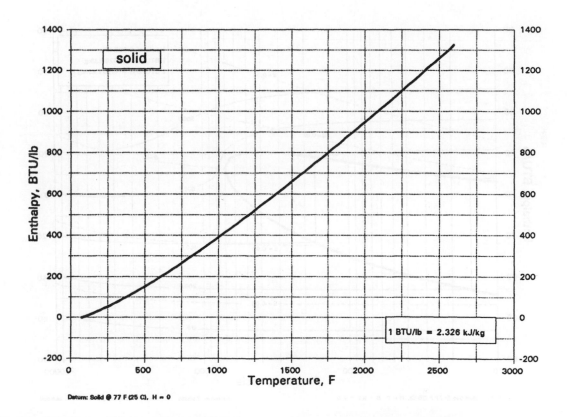

solid

1 BTU/lb = 2.326 kJ/kg

Datum: Solid @ 77 F (25 C), H = 0

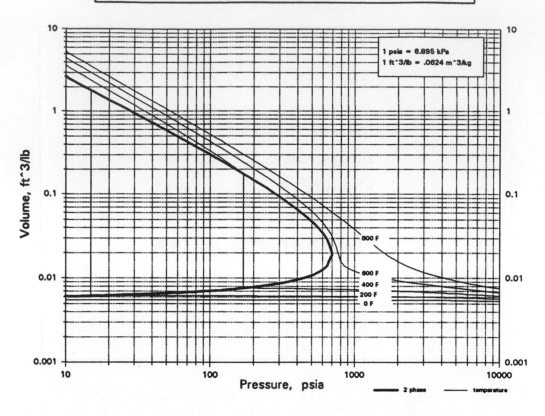

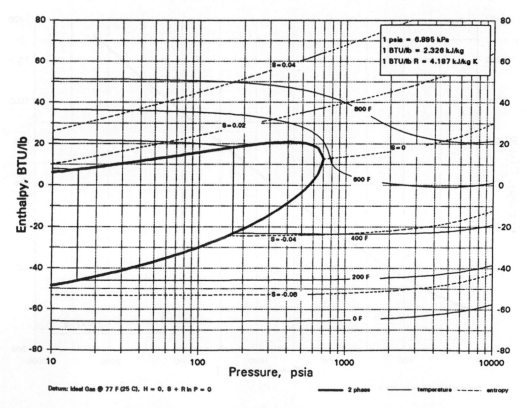

Datum: Ideal Gas @ 77 F (25 C), H = 0, S + R ln P = 0

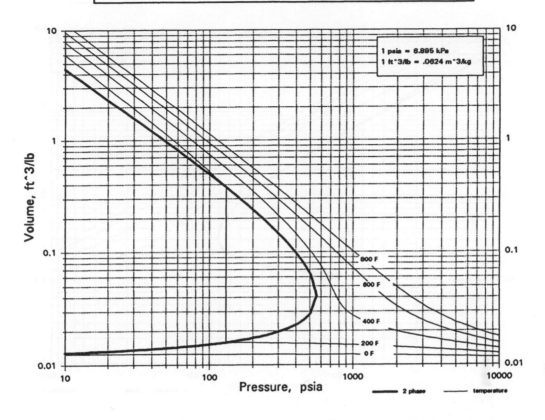

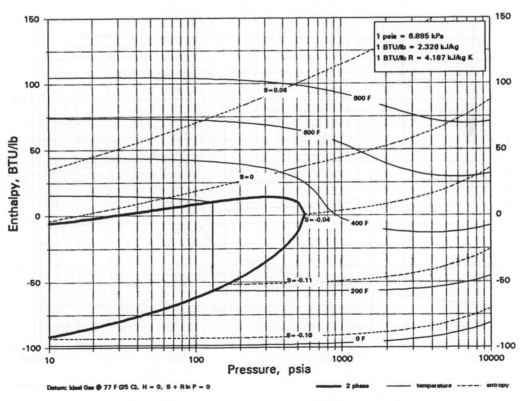

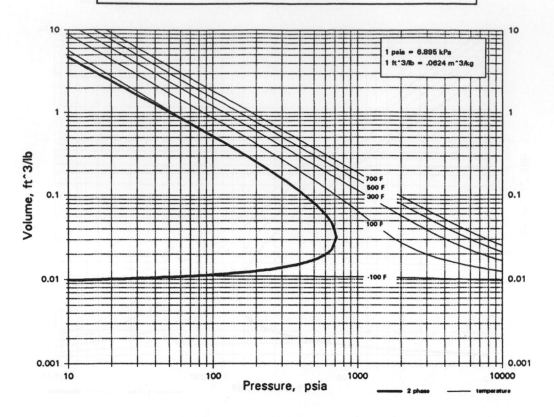

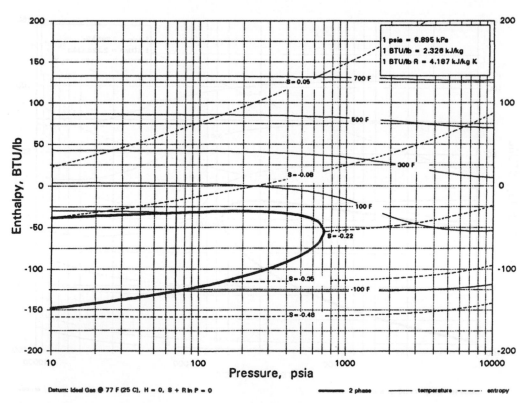

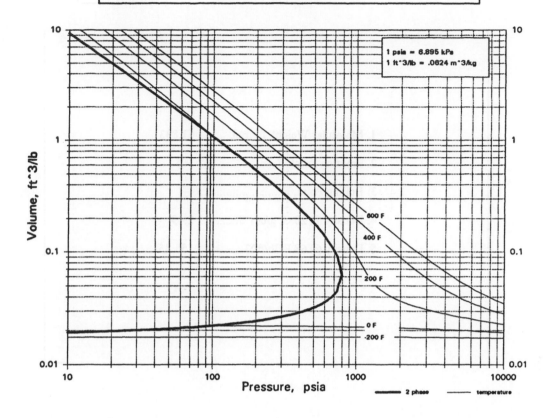

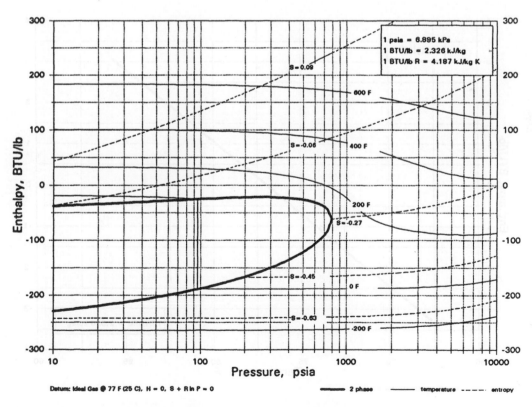

1. Molecular Weight, lb/mol.......... 61.833

2. Freezing Point, F....................... 365

3. Boiling Point, F......................... ---

4. Density @ 20 C, g/cm^3............ 2.49

5. Density @ 68 F, lb/ft^3............. 155.45

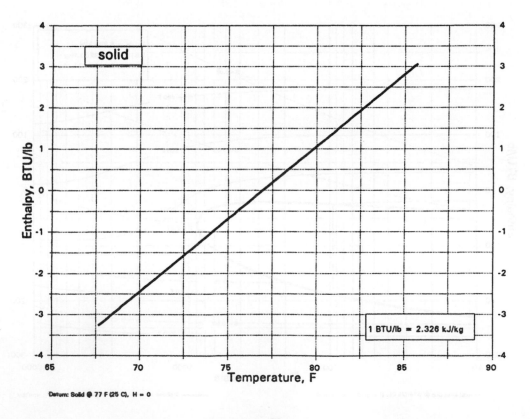

Datum: Solid @ 77 F (25 C), H = 0

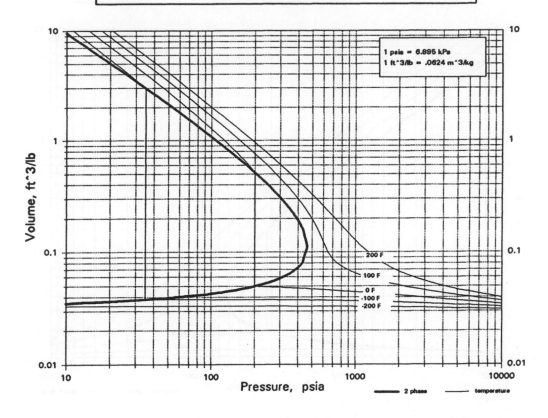

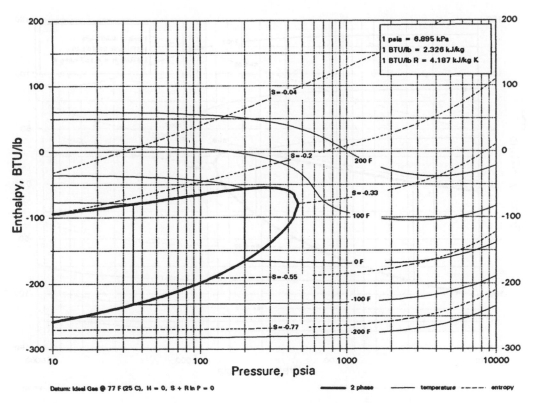

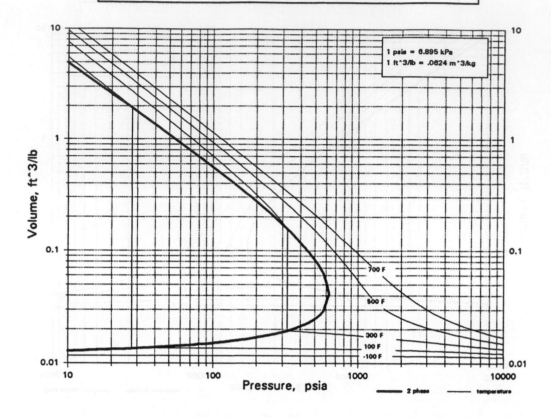

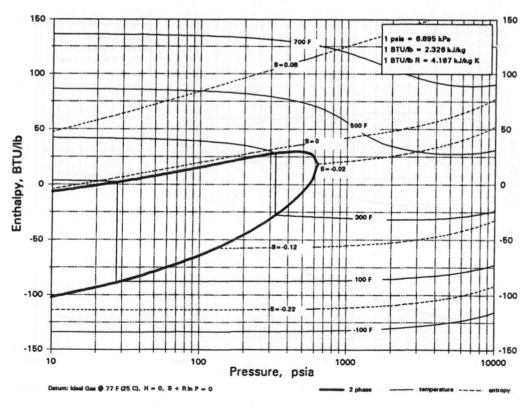

Datum: Ideal Gas @ 77 F (25 C), H = 0, S + R ln P = 0   2 phase   temperature   entropy

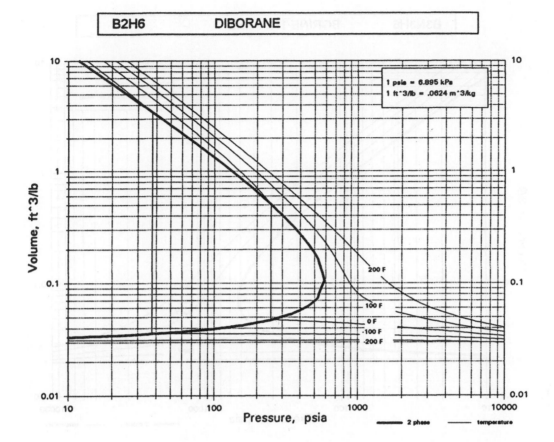

1 psia = 6.895 kPa
1 ft^3/lb = .0624 m^3/kg

200 F
100 F
0 F
-100 F
-200 F

2 phase   temperature

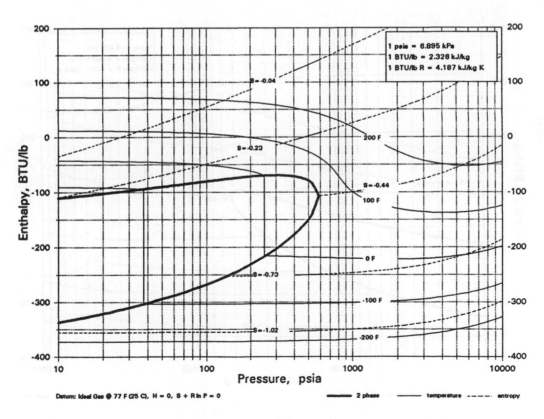

1 psia = 6.895 kPa
1 BTU/lb = 2.326 kJ/kg
1 BTU/lb R = 4.187 kJ/kg K

S = -0.04
S = -0.23
S = -0.44
S = -0.73
S = -1.02

200 F
100 F
0 F
-100 F
-200 F

Datum: Ideal Gas @ 77 F (25 C), H = 0, S + R ln P = 0

2 phase   temperature   ----- entropy

31

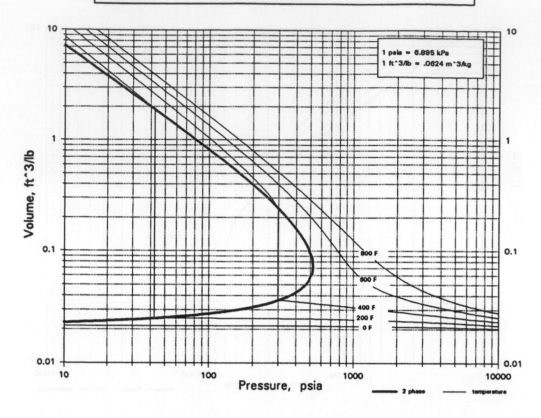

1 psia = 6.895 kPa
1 ft^3/lb = .0624 m^3/kg

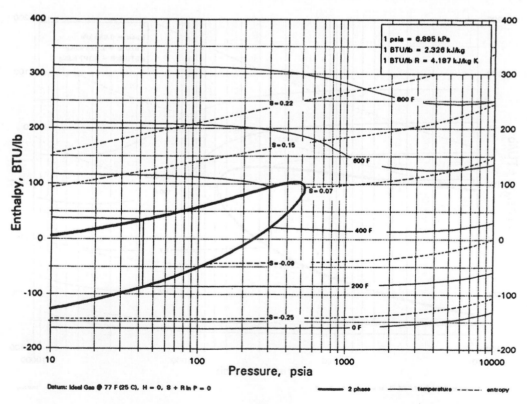

1 psia = 6.895 kPa
1 BTU/lb = 2.326 kJ/kg
1 BTU/lb R = 4.187 kJ/kg K

Datum: Ideal Gas @ 77 F (25 C), H = 0, S + R ln P = 0

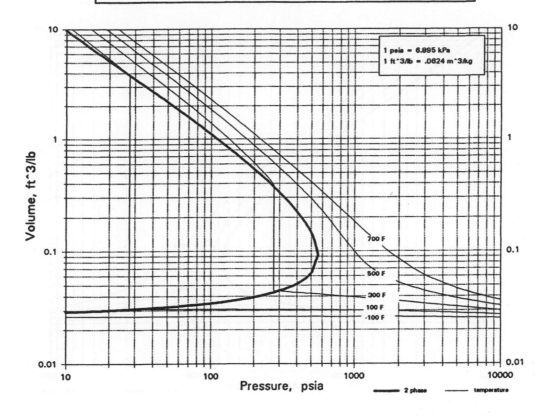

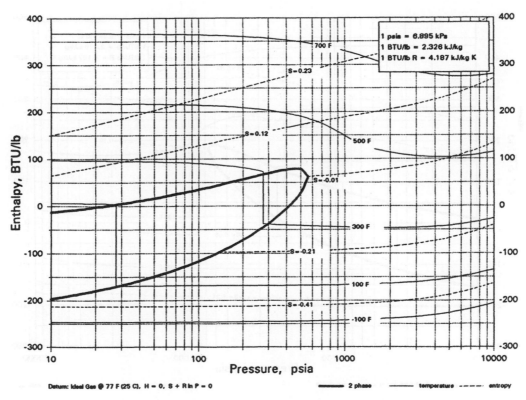

33

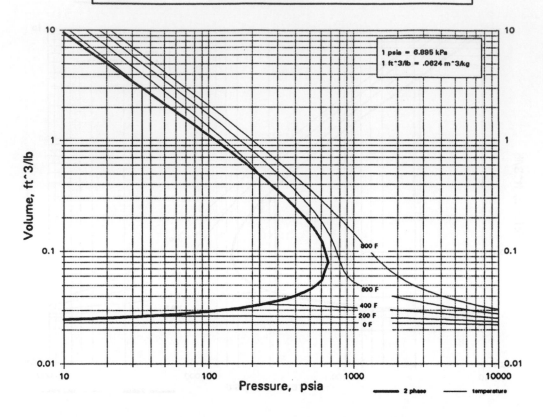

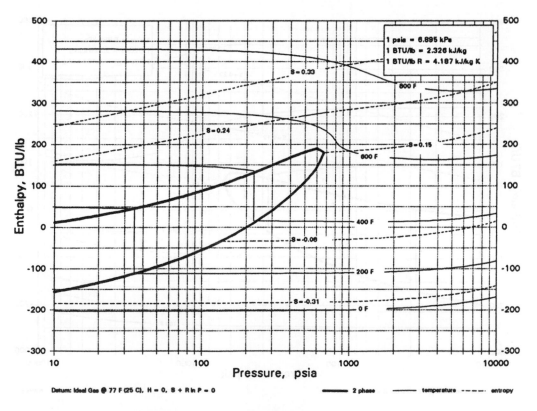

Datum: Ideal Gas @ 77 F (25 C), H = 0, S + R ln P = 0

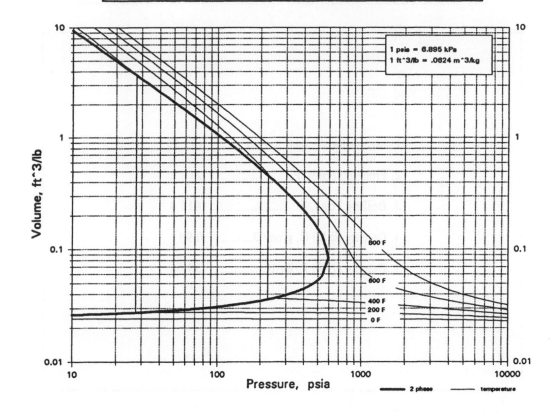

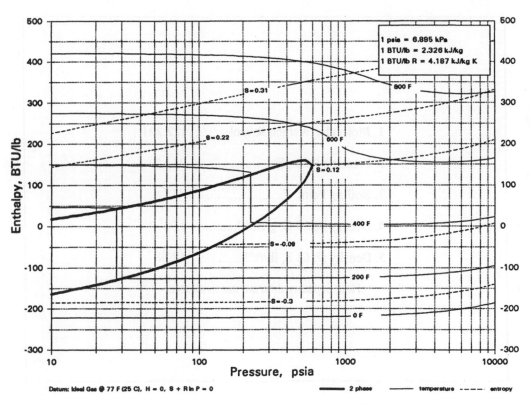

Datum: Ideal Gas @ 77 F (25 C), H = 0, S + R ln P = 0

1. Molecular Weight, lb/mol.......... 122.221

2. Freezing Point, F...................... 211.2

3. Boiling Point, F......................... 415.4

4. Density @ 20 C, g/cm^3............ 0.94

5. Density @ 68 F, lb/ft^3.............. 58.68

---

1. Molecular Weight, lb/mol.......... 122.221

2. Freezing Point, F...................... 211.2

3. Boiling Point, F......................... 415.4

4. Density @ 20 C, g/cm^3............ 0.94

5. Density @ 68 F, lb/ft^3.............. 58.68

Heat capacity data are not available.

1. Molecular Weight, lb/mol.......... 137.327

2. Freezing Point, F....................... 1340.6

3. Boiling Point, F......................... 2972.9

4. Density @ 20 C, g/cm^3............. 3.51

5. Density @ 68 F, lb/ft^3.............. 219.12

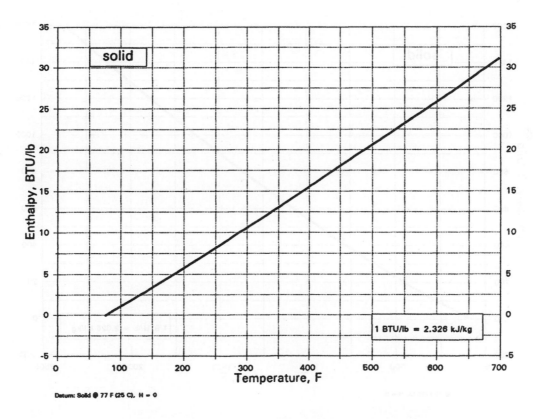

solid

1 BTU/lb = 2.326 kJ/kg

Enthalpy, BTU/lb

Temperature, F

Datum: Solid @ 77 F (25 C), H = 0

1. Molecular Weight, lb/mol.......... 9.012

2. Freezing Point, F....................... 2348.6

3. Boiling Point, F......................... 4479.5

4. Density @ 20 C, g/cm^3............ 1.85

5. Density @ 68 F, lb/ft^3.............. 115.49

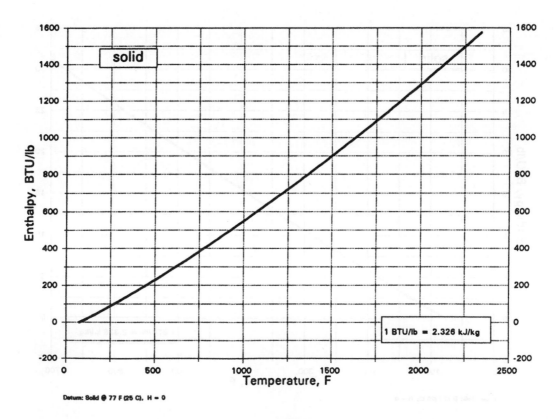

Datum: Solid @ 77 F (25 C), H = 0

1. Molecular Weight, lb/mol.......... 38.698

2. Freezing Point, F....................... 253.4

3. Boiling Point, F......................... 194

4. Density @ C, g/cm^3............ ---

5. Density @ F, lb/ft^3.............. ---

1. Molecular Weight, lb/mol.......... 38.698

2. Freezing Point, F....................... 253.4

3. Boiling Point, F......................... 194

4. Density @ C, g/cm^3............ ---

5. Density @ F, lb/ft^3.............. ---

Heat capacity data are not available.

1. Molecular Weight, lb/mol.......... 168.82

2. Freezing Point, F....................... 914

3. Boiling Point, F......................... 885.2

4. Density @ 25 C, g/cm^3............ 3.47

5. Density @ 77 F, lb/ft^3.............. 216.62

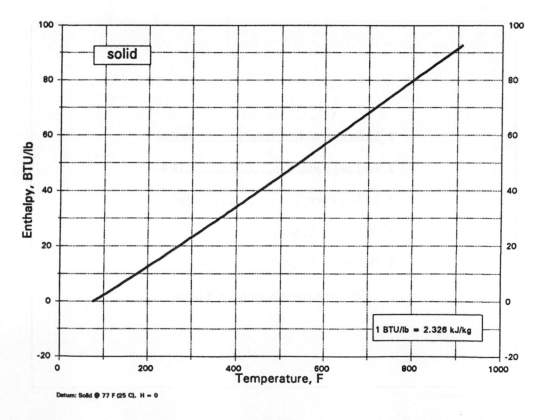

Datum: Solid @ 77 F (25 C). H = 0

40

1. Molecular Weight, lb/mol.......... 79.918

2. Freezing Point, F....................... 761

3. Boiling Point, F.......................... 908.6

4. Density @ 25 C, g/cm^3............. 1.9

5. Density @ 77 F, lb/ft^3.............. 118.61

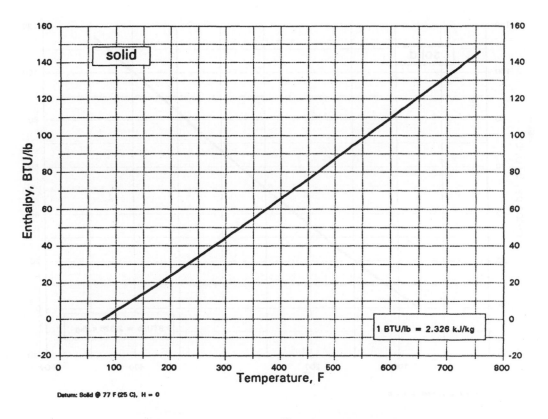

solid

1 BTU/lb = 2.326 kJ/kg

Enthalpy, BTU/lb

Temperature, F

Datum: Solid @ 77 F (25 C), H = 0

1. Molecular Weight, lb/mol.......... 47.009

2. Freezing Point, F....................... 1472

3. Boiling Point, F.......................... ---

4. Density @ 25 C, g/cm^3............. 1.99

5. Density @ 77 F, lb/ft^3.............. 124.23

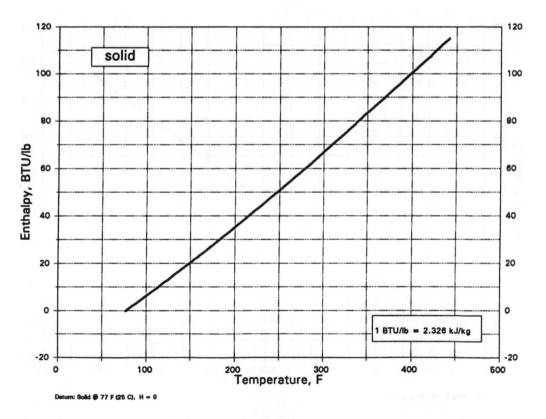

solid

1 BTU/lb = 2.326 kJ/kg

Enthalpy, BTU/lb

Temperature, F

Datum: Solid @ 77 F (25 C), H = 0

1. Molecular Weight, lb/mol.......... 262.821

2. Freezing Point, F........................ 910.4

3. Boiling Point, F.......................... 908.6

4. Density @ 26 C, g/cm^3............. 4.33

5. Density @ 79 F, lb/ft^3.............. 270.31

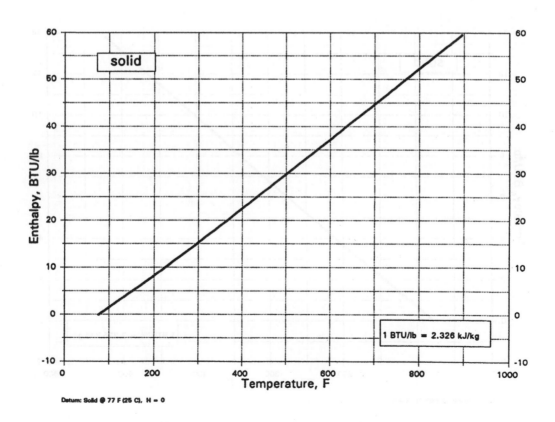

Datum: Solid @ 77 F (25 C), H = 0

1. Molecular Weight, lb/mol.......... 208.98

2. Freezing Point, F....................... 519.8

3. Boiling Point, F......................... 2597

4. Density @ 20 C, g/cm^3............ 9.8

5. Density @ 68 F, lb/ft^3.............. 611.79

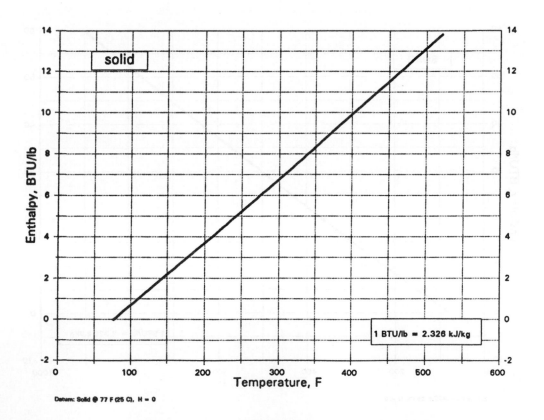

Datum: Solid @ 77 F (25 C), H = 0

1. Molecular Weight, lb/mol.......... 448.692

2. Freezing Point, F........................ 424.4

3. Boiling Point, F.......................... 861.8

4. Density @ 25 C, g/cm^3............ 5.72

5. Density @ 77 F, lb/ft^3.............. 357.09

---

1. Molecular Weight, lb/mol.......... 448.692

2. Freezing Point, F........................ 424.4

3. Boiling Point, F.......................... 861.8

4. Density @ 25 C, g/cm^3............ 5.72

5. Density @ 77 F, lb/ft^3.............. 357.09

Heat capacity data are not available.

1. Molecular Weight, lb/mol.......... 315.338

2. Freezing Point, F........................ 446

3. Boiling Point, F.......................... 825.8

4. Density @ 25 C, g/cm^3............. 4.75

5. Density @ 77 F, lb/ft^3.............. 296.53

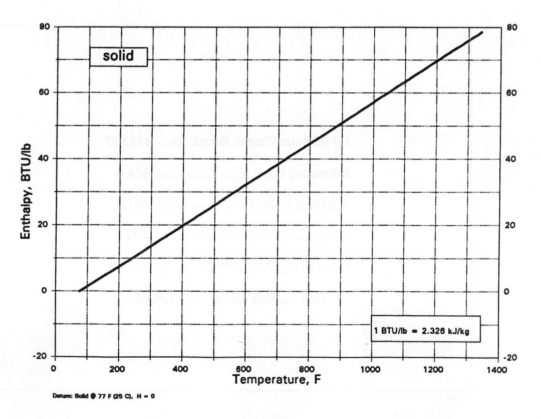

Datum: Solid @ 77 F (25 C), H = 0

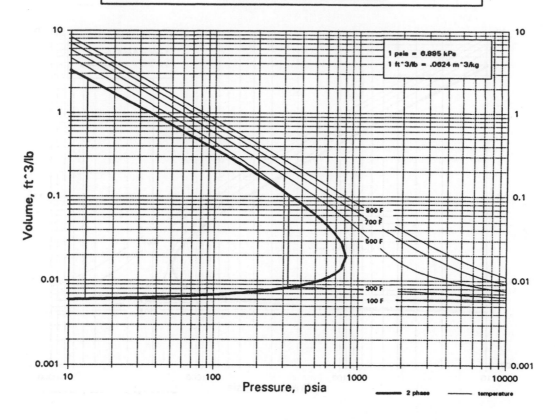

1 psia = 6.895 kPa
1 ft^3/lb = .0624 m^3/kg

900 F
700 F
500 F
300 F
100 F

Volume, ft^3/lb

Pressure, psia

2 phase   temperature

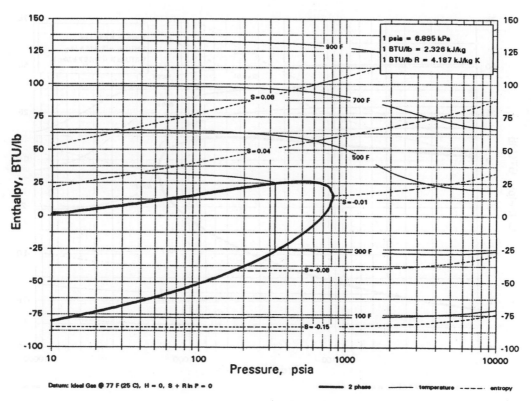

1 psia = 6.895 kPa
1 BTU/lb = 2.326 kJ/kg
1 BTU/lb R = 4.187 kJ/kg K

900 F
S = 0.08
700 F
S = 0.04
500 F
S = -0.01
300 F
S = -0.08
100 F
S = -0.15

Enthalpy, BTU/lb

Pressure, psia

Datum: Ideal Gas @ 77 F (25 C), H = 0, S + R ln P = 0

2 phase   temperature   entropy

47

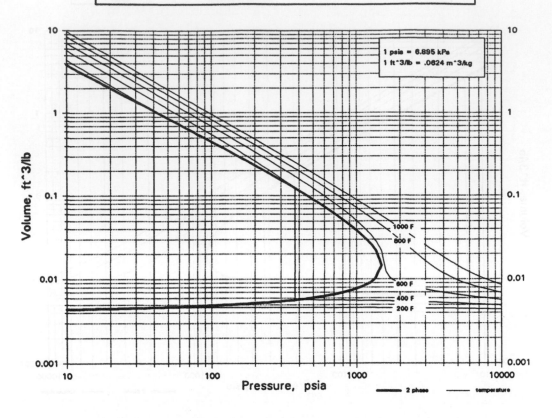

1 psia = 6.895 kPa
1 ft^3/lb = .0624 m^3/kg

1000 F
800 F
600 F
400 F
200 F

Volume, ft^3/lb

Pressure, psia

2 phase   temperature

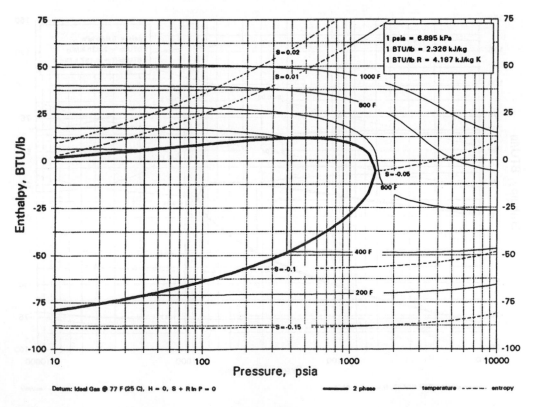

1 psia = 6.895 kPa
1 BTU/lb = 2.326 kJ/kg
1 BTU/lb R = 4.187 kJ/kg K

S = 0.02
S = 0.01
1000 F
800 F
S = -0.05
600 F
400 F
S = -0.1
200 F
S = -0.15

Enthalpy, BTU/lb

Pressure, psia

Datum: Ideal Gas @ 77 F (25 C), H = 0, S + R ln P = 0

2 phase   temperature   - - - - entropy

1. Molecular Weight, lb/mol.......... 12.011

2. Freezing Point, F........................ 7184.9

3. Boiling Point, F......................... 7105.7

4. Density @ 20 C, g/cm^3............ 2.25

5. Density @ 68 F, lb/ft^3.............. 140.46

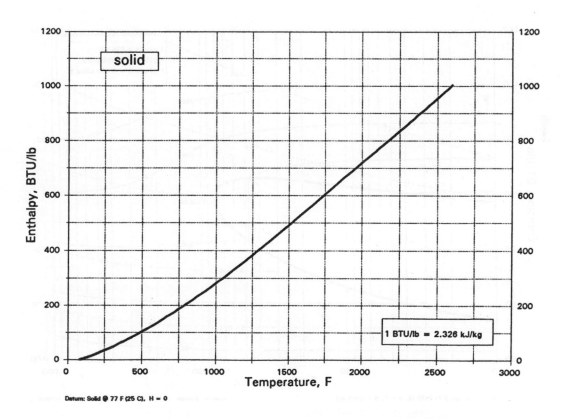

Datum: Solid @ 77 F (25 C), H = 0

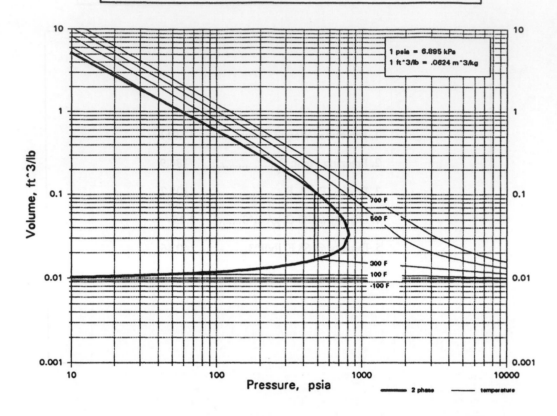

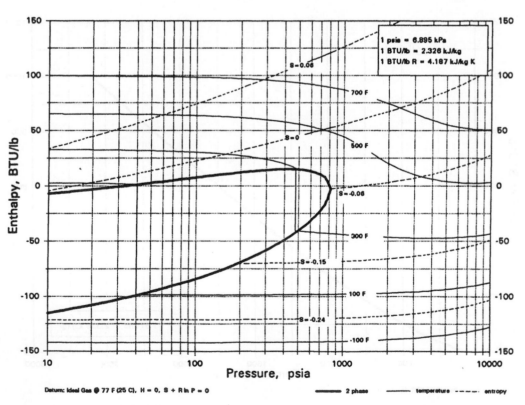

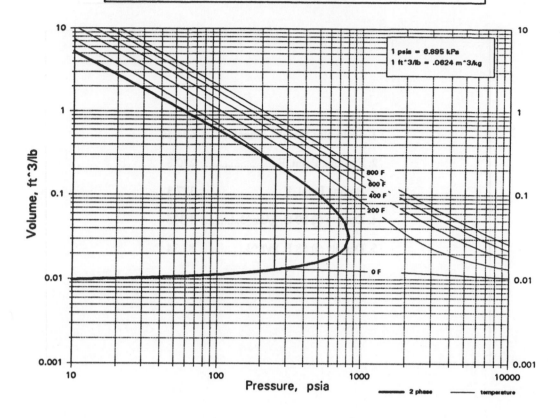

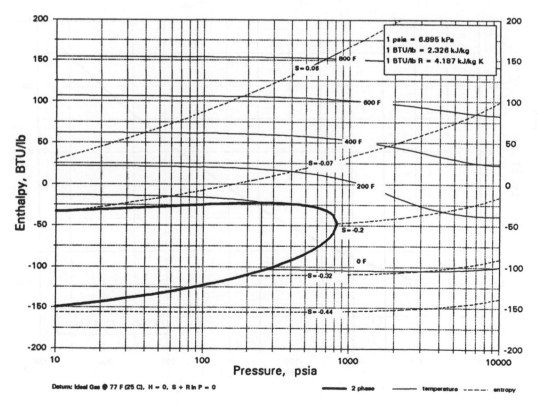

1. Molecular Weight, lb/mol.......... 60.056

2. Freezing Point, F..................... 270.8

3. Boiling Point, F....................... 377.3

4. Density @ 20 C, g/cm^3............ 1.335

5. Density @ 68 F, lb/ft^3.............. 83.34

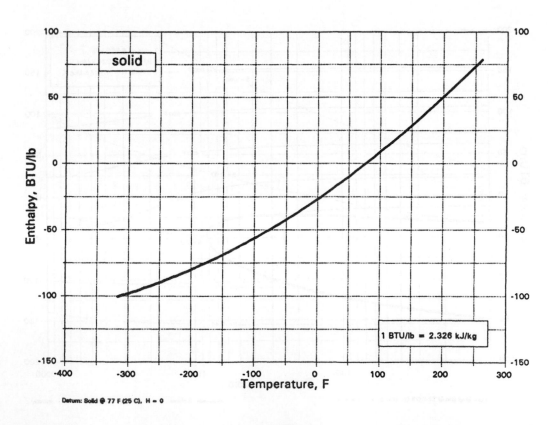

solid

1 BTU/lb = 2.326 kJ/kg

Datum: Solid @ 77 F (25 C), H = 0

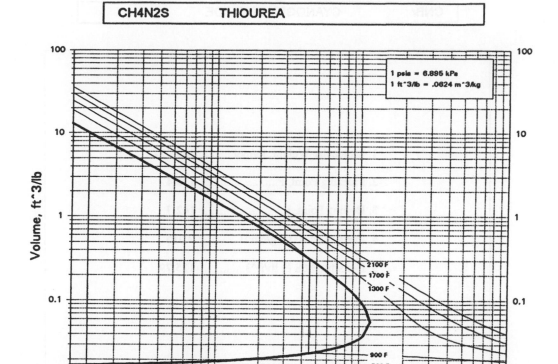

1 psia = 6.895 kPa
1 ft^3/lb = .0624 m^3/kg

2100 F
1700 F
1300 F

900 F
500 F

Volume, ft^3/lb

Pressure, psia

2 phase   temperature

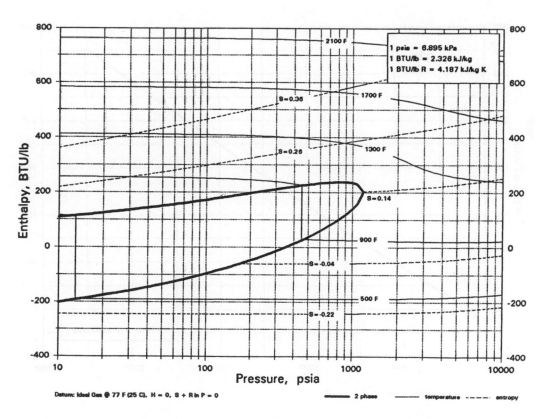

1 psia = 6.895 kPa
1 BTU/lb = 2.326 kJ/kg
1 BTU/lb R = 4.187 kJ/kg K

2100 F

S=0.36   1700 F

S=0.26   1300 F

S=0.14

900 F

S=-0.04

500 F

S=-0.22

Enthalpy, BTU/lb

Pressure, psia

Datum: Ideal Gas @ 77 F (25 C), H = 0, S + R ln P = 0

2 phase   temperature   entropy

53

CNBr   CYANOGEN BROMIDE

1. Molecular Weight, lb/mol.......... 105.922

2. Freezing Point, F........................ 136.4

3. Boiling Point, F......................... 142.7

4. Density @ 20 C, g/cm^3............. 2.015

5. Density @ 68 F, lb/ft^3.............. 125.79

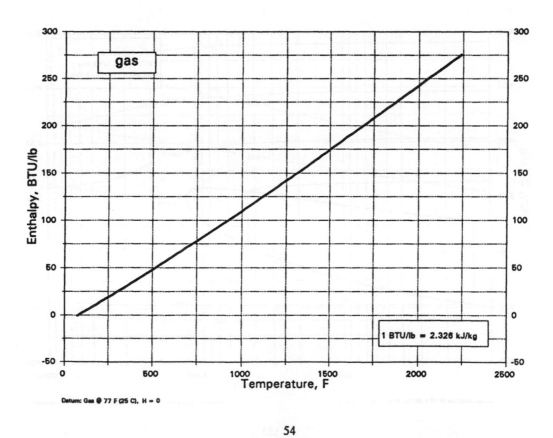

gas

1 BTU/lb = 2.326 kJ/kg

Datum: Gas @ 77 F (25 C), H = 0

CNCl — CYANOGEN CHLORIDE

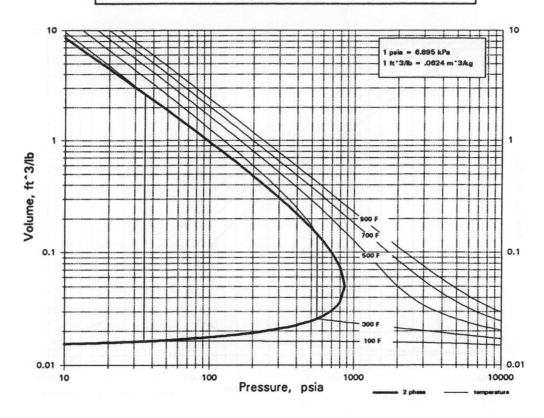

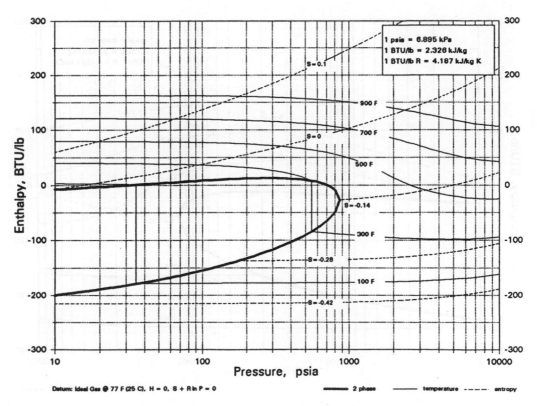

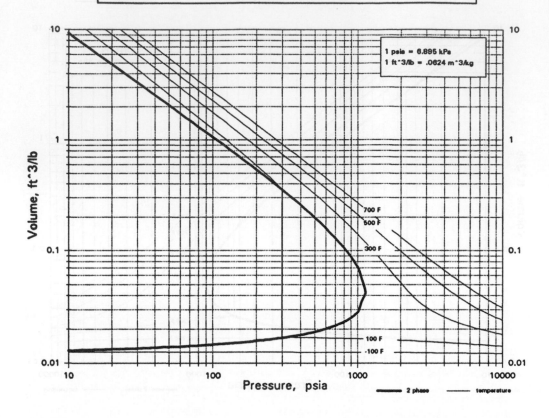

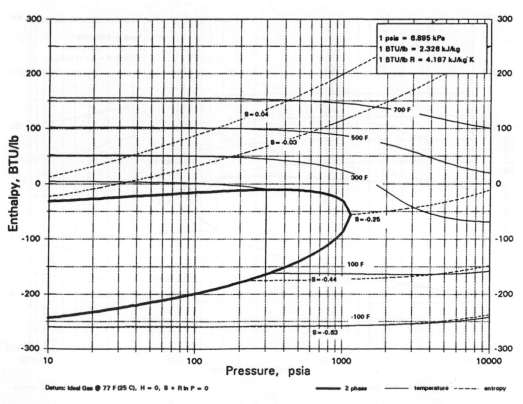

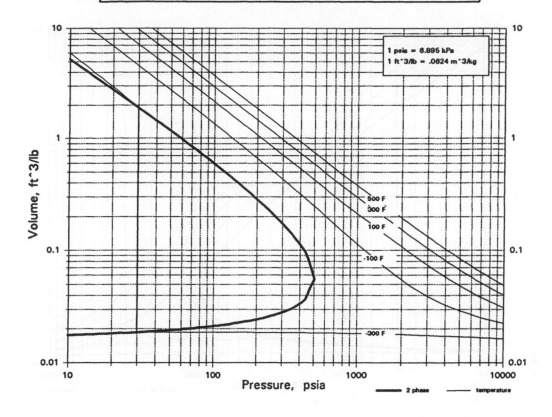

1 psia = 6.895 kPa
1 ft^3/lb = .0624 m^3/kg

Volume, ft^3/lb

500 F
300 F
100 F
-100 F
-300 F

Pressure, psia

2 phase   temperature

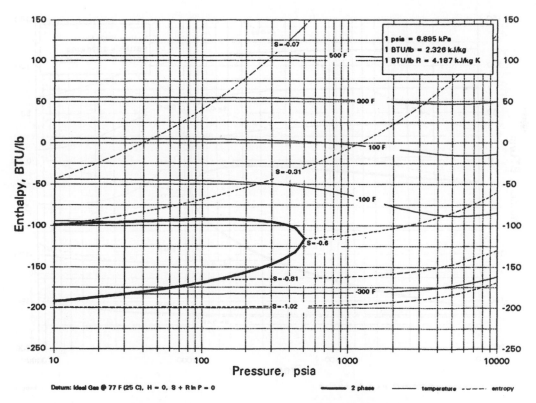

1 psia = 6.895 kPa
1 BTU/lb = 2.326 kJ/kg
1 BTU/lb R = 4.187 kJ/kg K

Enthalpy, BTU/lb

S = -0.07
500 F
300 F
100 F
S = -0.31
-100 F
S = -0.6
S = -0.81
-300 F
S = -1.02

Pressure, psia

Datum: Ideal Gas @ 77 F (25 C), H = 0, S + R ln P = 0

2 phase   temperature   ----- entropy

57

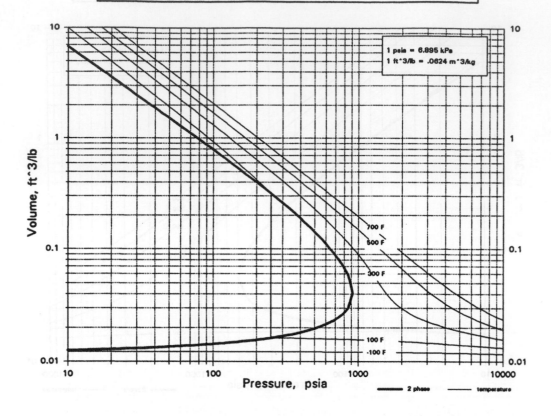

1 psia = 6.895 kPa
1 ft^3/lb = .0624 m^3/kg

700 F
500 F
300 F
100 F
-100 F

2 phase   temperature

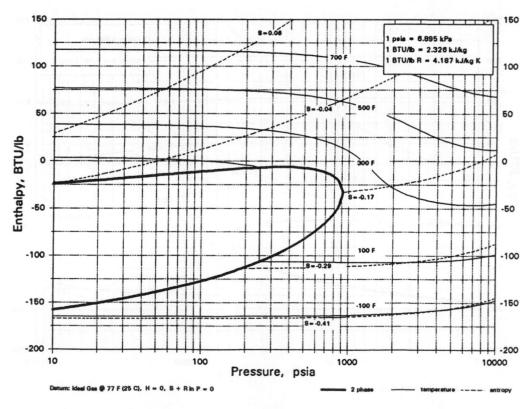

1 psia = 6.895 kPa
1 BTU/lb = 2.326 kJ/kg
1 BTU/lb R = 4.187 kJ/kg K

S=0.08
S=-0.04
700 F
500 F
300 F
S=-0.17
100 F
S=-0.29
-100 F
S=-0.41

Datum: Ideal Gas @ 77 F (25 C), H = 0, S + R ln P = 0

2 phase   temperature   entropy

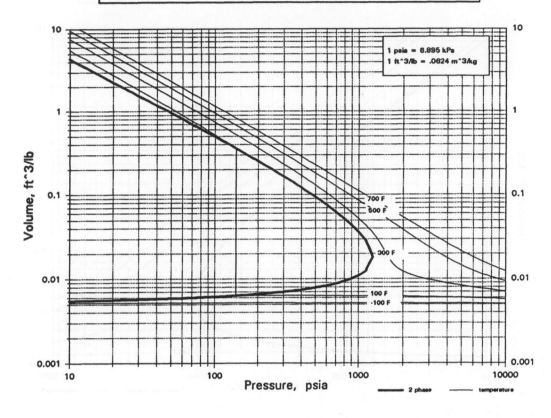

1 psia = 6.895 kPa
1 ft^3/lb = .0624 m^3/kg

700 F
500 F
300 F
100 F
-100 F

Volume, ft^3/lb

Pressure, psia

— 2 phase   — temperature

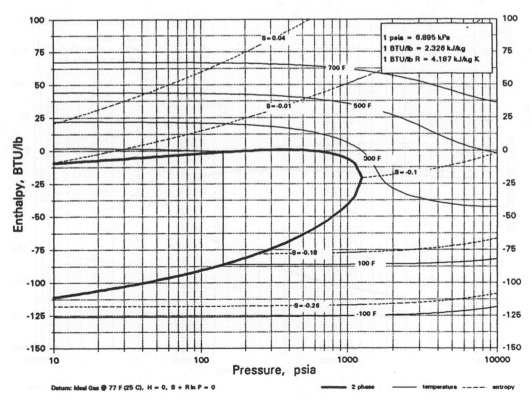

1 psia = 6.895 kPa
1 BTU/lb = 2.326 kJ/kg
1 BTU/lb R = 4.187 kJ/kg K

S = 0.04
S = -0.01
S = -0.1
S = -0.18
S = -0.26

700 F
500 F
300 F
100 F
-100 F

Enthalpy, BTU/lb

Pressure, psia

Datum: Ideal Gas @ 77 F (25 C), H = 0, S + R ln P = 0

— 2 phase   — temperature   ---- entropy

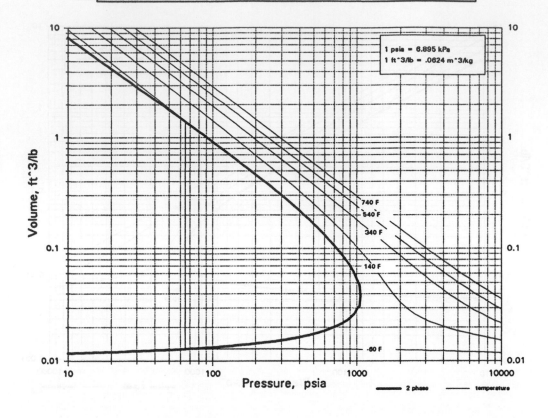

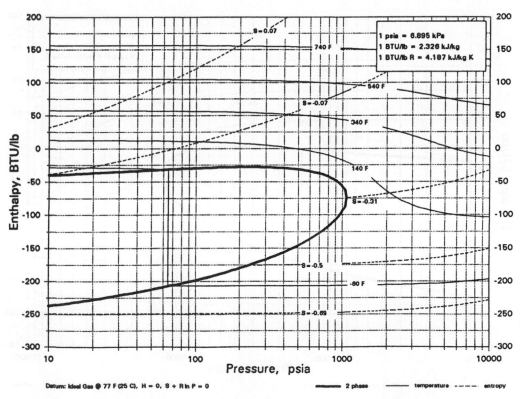

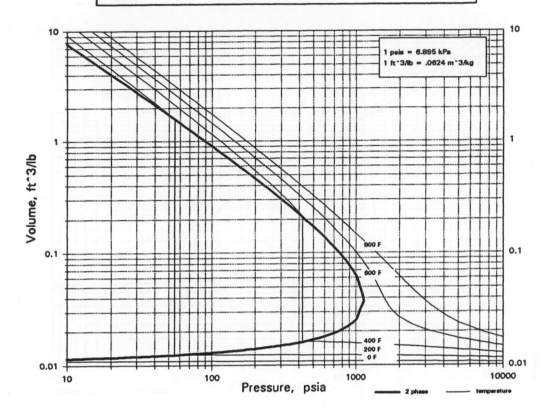

1 psia = 6.895 kPa
1 ft^3/lb = .0624 m^3/kg

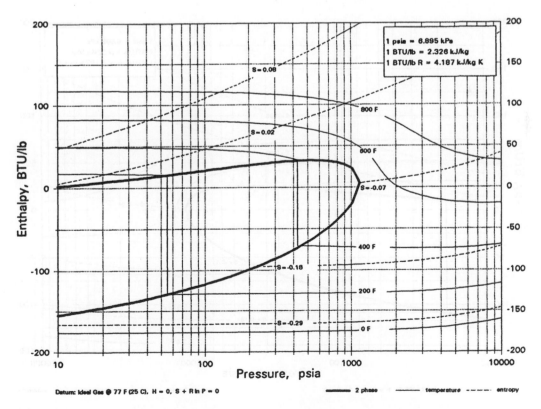

1 psia = 6.895 kPa
1 BTU/lb = 2.326 kJ/kg
1 BTU/lb R = 4.187 kJ/kg K

Datum: Ideal Gas @ 77 F (25 C), H = 0, S + R ln P = 0

61

CSeS   CARBON SELENOSULFIDE

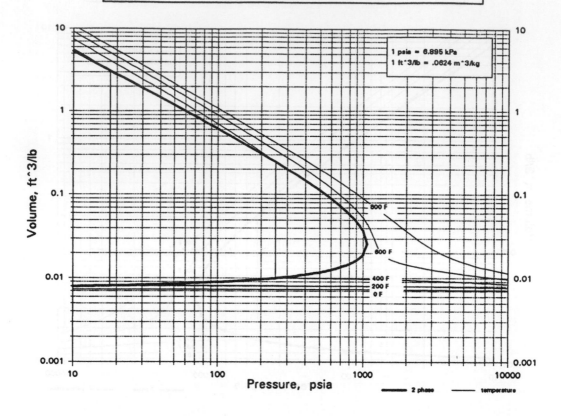

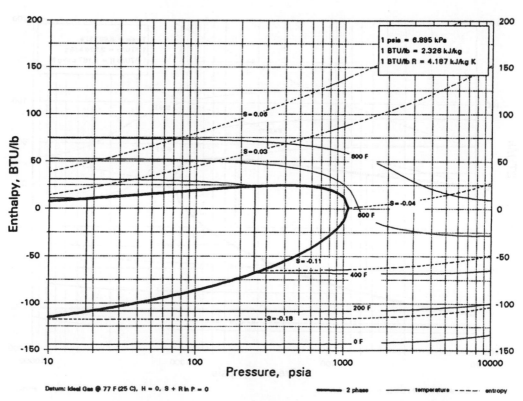

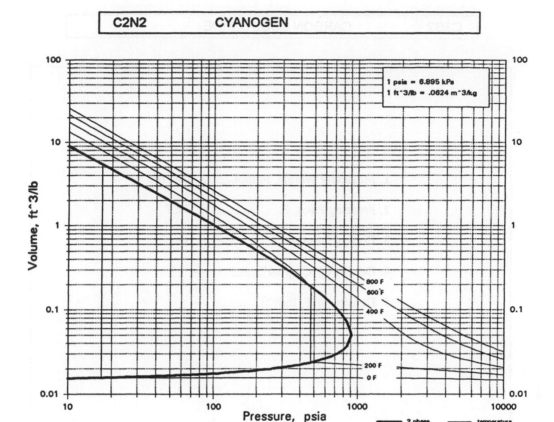

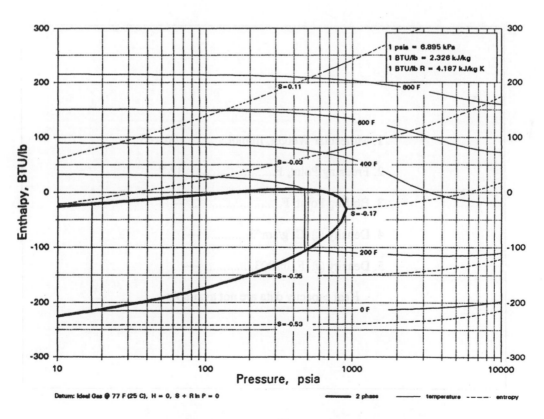

Datum: Ideal Gas @ 77 F (25 C), H = 0, S + R ln P = 0

63

1. Molecular Weight, lb/mol.......... 100.165

2. Freezing Point, F....................... 32.7

3. Boiling Point, F......................... ---

4. Density @ C, g/cm^3............. ---

5. Density @ F, lb/ft^3.............. ---

1. Molecular Weight, lb/mol.......... 100.165

2. Freezing Point, F....................... 32.7

3. Boiling Point, F......................... ---

4. Density @ C, g/cm^3............. ---

5. Density @ F, lb/ft^3.............. ---

Heat capacity data are not available.

Ca	CALCIUM

1. Molecular Weight, lb/mol.......... 40.078

2. Freezing Point, F...................... 1547.6

3. Boiling Point, F......................... 2711.9

4. Density @ 20 C, g/cm^3............. 1.54

5. Density @ 68 F, lb/ft^3.............. 96.14

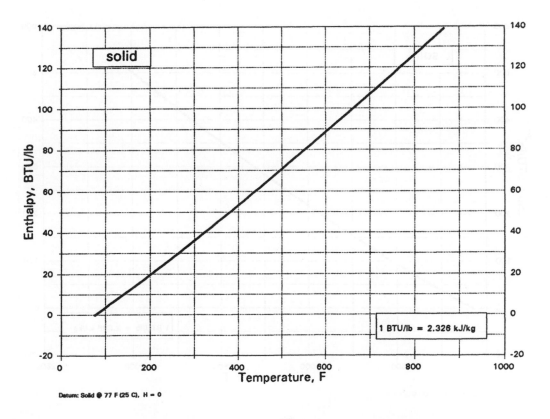

solid

1 BTU/lb = 2.326 kJ/kg

Datum: Solid @ 77 F (25 C), H = 0

1. Molecular Weight, lb/mol.......... 78.075

2. Freezing Point, F........................ 2584.1

3. Boiling Point, F......................... 4592

4. Density @ 20 C, g/cm^3............ 3.18

5. Density @ 68 F, lb/ft^3.............. 198.52

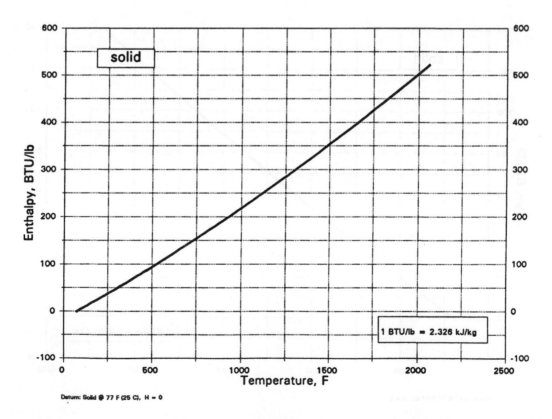

Datum: Solid @ 77 F (25 C), H = 0

1. Molecular Weight, lb/mol........... 187.898

2. Freezing Point, F....................... 167.9

3. Boiling Point, F.......................... 437

4. Density @ 20 C, g/cm^3............ 3.29

5. Density @ 68 F, lb/ft^3.............. 205.39

1. Molecular Weight, lb/mol.......... 187.898

2. Freezing Point, F....................... 167.9

3. Boiling Point, F.......................... 437

4. Density @ 20 C, g/cm^3............ 3.29

5. Density @ 68 F, lb/ft^3.............. 205.39

Heat capacity data are not available.

1. Molecular Weight, lb/mol.......... 112.411

2. Freezing Point, F...................... 609.6

3. Boiling Point, F........................ 1418

4. Density @ 20 C, g/cm^3............. 8.64

5. Density @ 68 F, lb/ft^3.............. 539.38

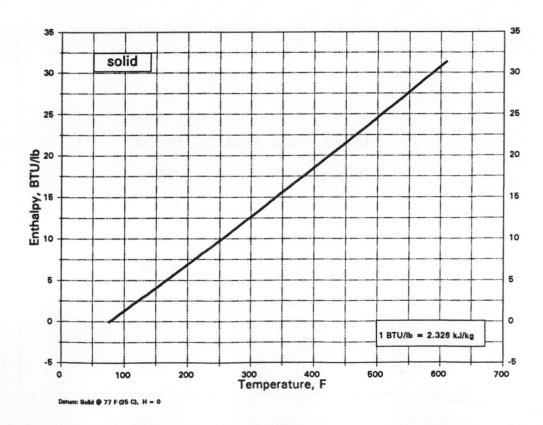

Datum: Solid @ 77 F (25 C), H = 0

1. Molecular Weight, lb/mol.......... 183.316

2. Freezing Point, F........................ 1054.4

3. Boiling Point, F......................... 1772.6

4. Density @ 25 C, g/cm^3............. 4.05

5. Density @ 77 F, lb/ft^3.............. 252.83

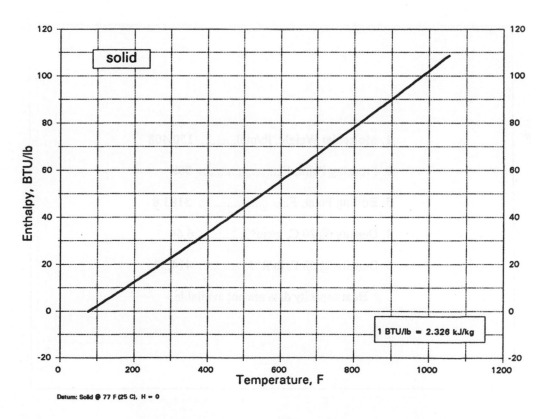

Datum: Solid @ 77 F (25 C), H = 0

1 BTU/lb = 2.326 kJ/kg

1. Molecular Weight, lb/mol.......... 150.408

2. Freezing Point, F....................... 968

3. Boiling Point, F......................... 3183.8

4. Density @ 20 C, g/cm^3............. 6.64

5. Density @ 68 F, lb/ft^3.............. 414.52

1. Molecular Weight, lb/mol.......... 150.408

2. Freezing Point, F....................... 968

3. Boiling Point, F......................... 3183.8

4. Density @ 20 C, g/cm^3............. 6.64

5. Density @ 68 F, lb/ft^3.............. 414.52

Heat capacity data are not available.

1. Molecular Weight, lb/mol.......... 366.22

2. Freezing Point, F....................... 725

3. Boiling Point, F.......................... 1464.8

4. Density @ 30 C, g/cm^3............. 5.67

5. Density @ 86 F, lb/ft^3.............. 353.97

---

1. Molecular Weight, lb/mol.......... 366.22

2. Freezing Point, F....................... 725

3. Boiling Point, F.......................... 1464.8

4. Density @ 30 C, g/cm^3............. 5.67

5. Density @ 86 F, lb/ft^3.............. 353.97

Heat capacity data are not available.

1. Molecular Weight, lb/mol.......... 128.41

2. Freezing Point, F...................... ---

3. Boiling Point, F......................... 2838.2

4. Density @ 20 C, g/cm^3............ 6.95

5. Density @ 68 F, lb/ft^3.............. 433.87

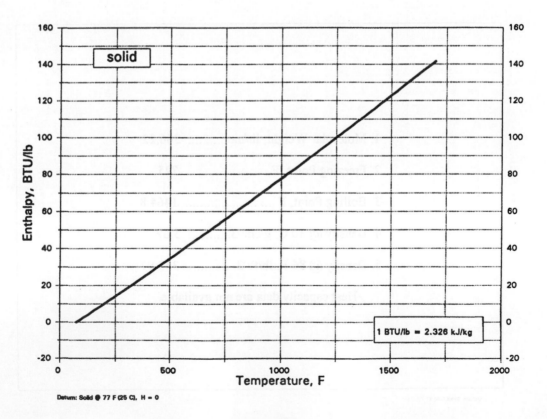

solid

Enthalpy, BTU/lb

Temperature, F

1 BTU/lb = 2.326 kJ/kg

Datum: Solid @ 77 F (25 C), H = 0

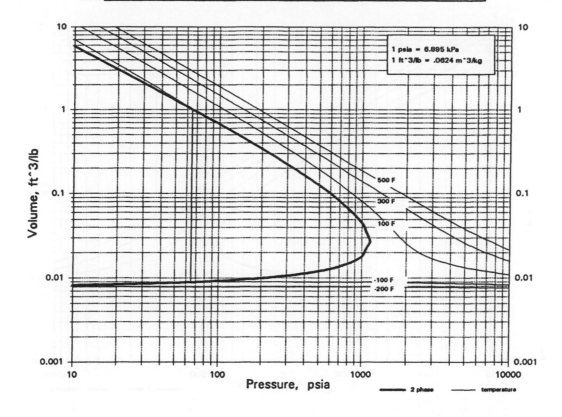

1 psia = 6.895 kPa
1 ft^3/lb = .0624 m^3/kg

—— 2 phase —— temperature

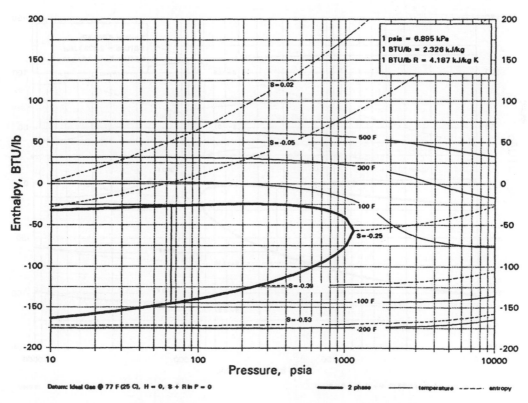

1 psia = 6.895 kPa
1 BTU/lb = 2.326 kJ/kg
1 BTU/lb R = 4.187 kJ/kg K

Datum: Ideal Gas @ 77 F (25 C), H = 0, S + R ln P = 0

—— 2 phase —— temperature ---- entropy

73

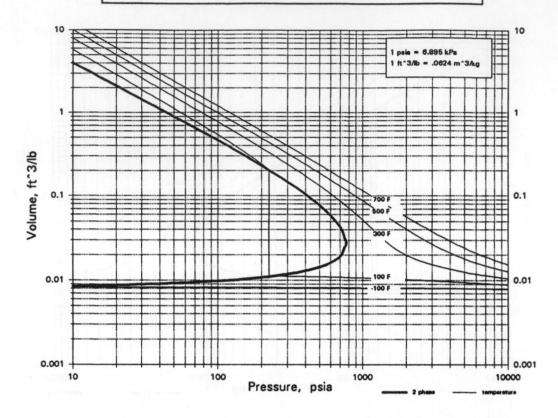

1 psia = 6.895 kPa
1 ft^3/lb = .0624 m^3/kg

700 F
500 F
300 F
100 F
-100 F

Volume, ft^3/lb

Pressure, psia

2 phase   temperature

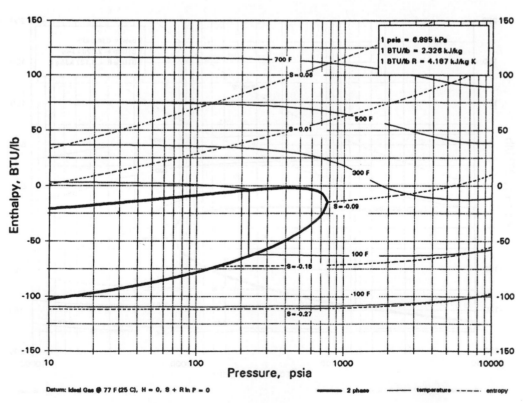

1 psia = 6.895 kPa
1 BTU/lb = 2.326 kJ/kg
1 BTU/lb R = 4.187 kJ/kg K

700 F
S=0.06
S=0.01
500 F
300 F
S=-0.09
100 F
S=-0.18
-100 F
S=-0.27

Enthalpy, BTU/lb

Pressure, psia

Datum: Ideal Gas @ 77 F (25 C), H = 0, S + R ln P = 0

2 phase   temperature ----- entropy

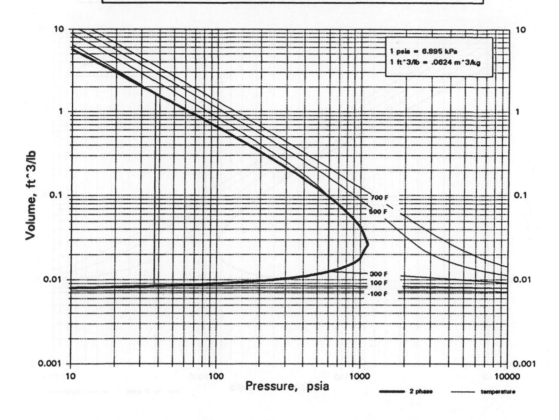

1 psia = 6.895 kPa
1 ft^3/lb = .0624 m^3/kg

700 F
500 F

300 F
100 F
-100 F

Volume, ft^3/lb

Pressure, psia

2 phase   temperature

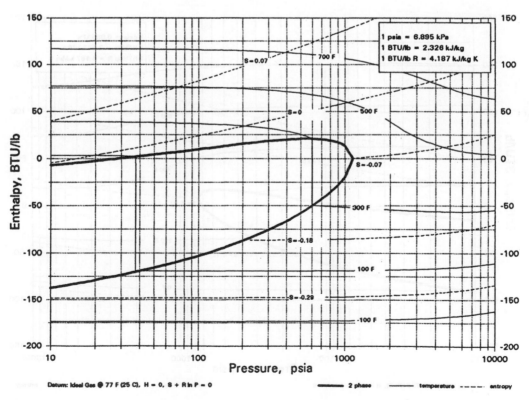

1 psia = 6.895 kPa
1 BTU/lb = 2.326 kJ/kg
1 BTU/lb R = 4.187 kJ/kg K

S=0.07   700 F

S=0   500 F

S=-0.07

300 F

S=-0.18

100 F

S=-0.29

-100 F

Enthalpy, BTU/lb

Pressure, psia

Datum: Ideal Gas @ 77 F (25 C), H = 0, S + R ln P = 0   2 phase   temperature ----- entropy

75

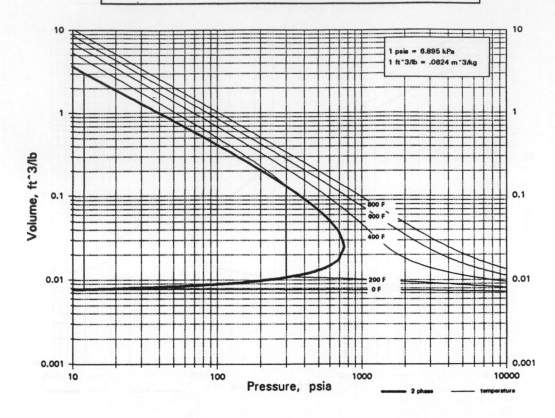

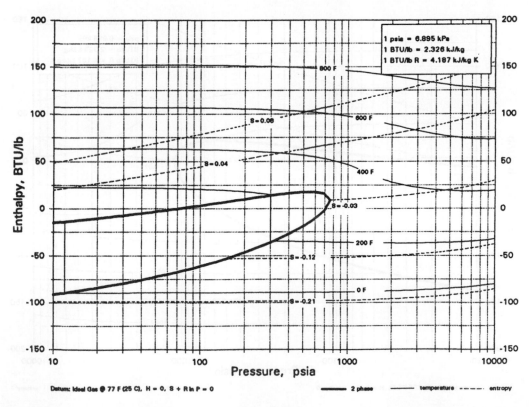

CIHO3S   CHLOROSULFONIC ACID

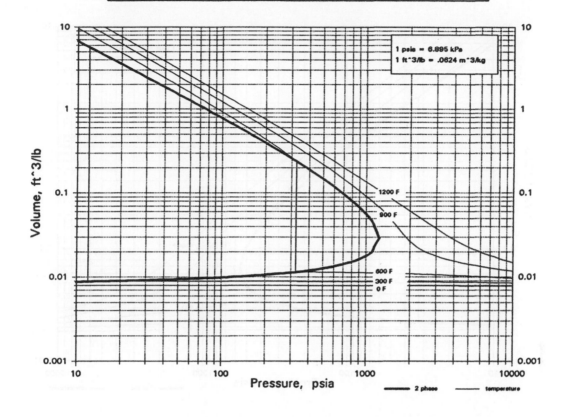

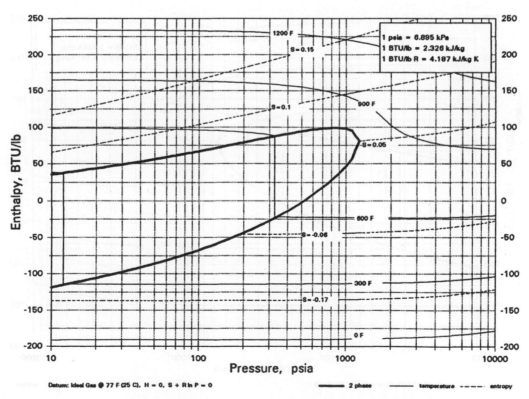

Datum: Ideal Gas @ 77 F (25 C), H = 0, S + R ln P = 0

77

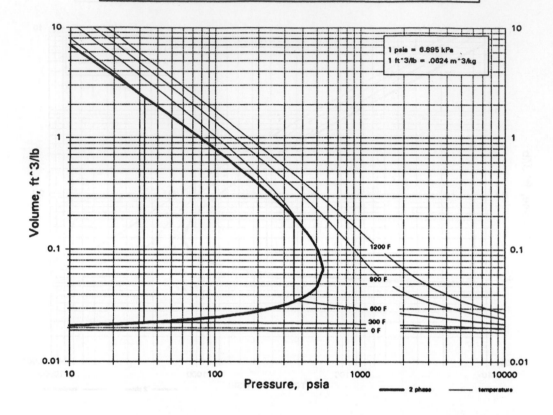

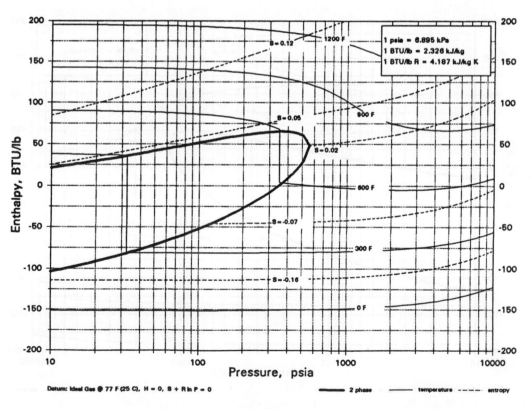

CIO2 CHLORINE DIOXIDE

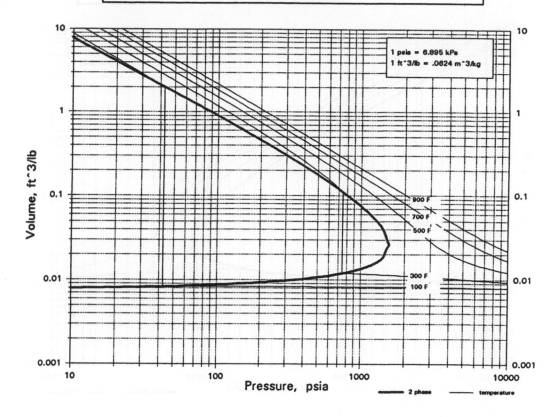

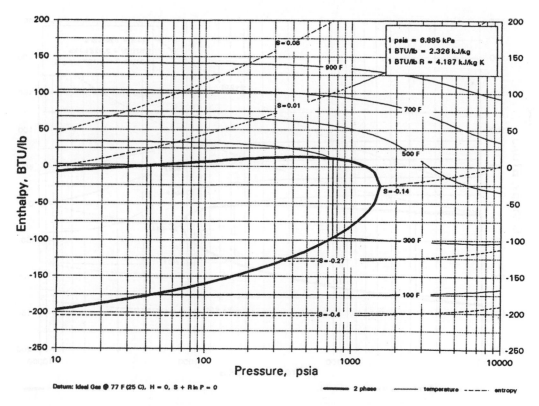

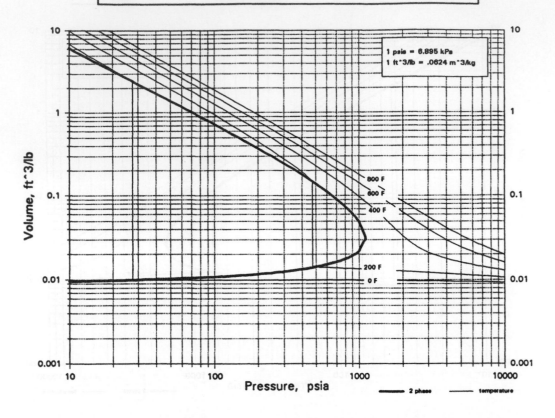

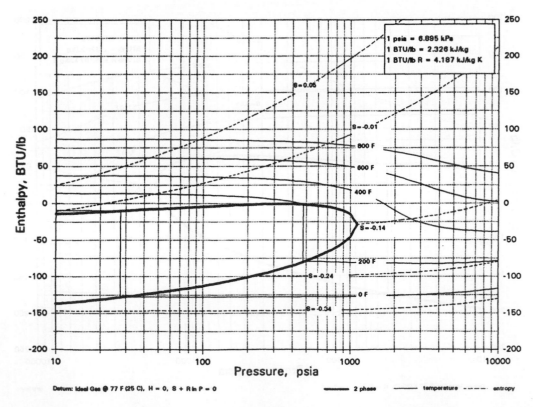

Datum: Ideal Gas @ 77 F (25 C), H = 0, S + R ln P = 0

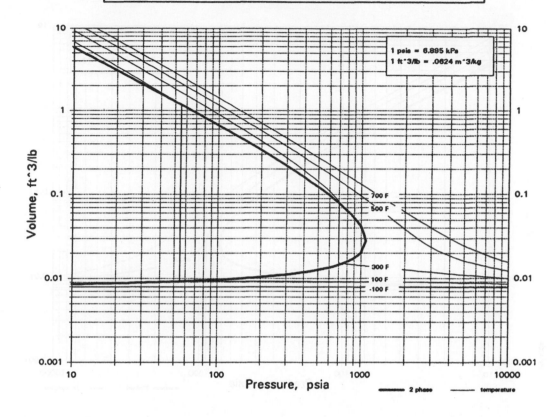

1 psia = 6.895 kPa
1 ft^3/lb = .0624 m^3/kg

700 F
500 F
300 F
100 F
-100 F

Volume, ft^3/lb

Pressure, psia

2 phase   temperature

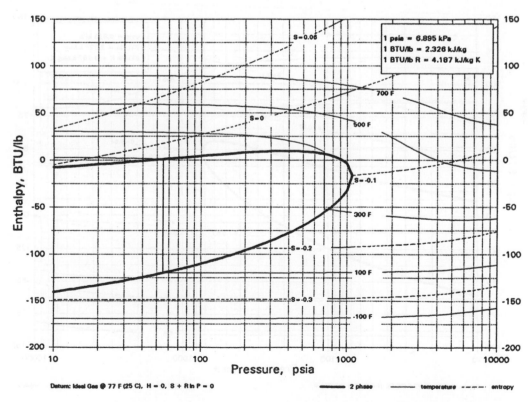

1 psia = 6.895 kPa
1 BTU/lb = 2.326 kJ/kg
1 BTU/lb R = 4.187 kJ/kg K

S=0.05
S=0
S=-0.1
S=-0.2
S=-0.3

700 F
500 F
300 F
100 F
-100 F

Enthalpy, BTU/lb

Pressure, psia

Datum: Ideal Gas @ 77 F (25 C), H = 0, S + R ln P = 0   2 phase   temperature ----- entropy

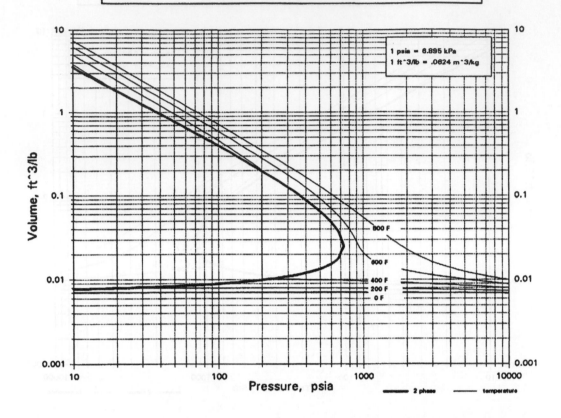

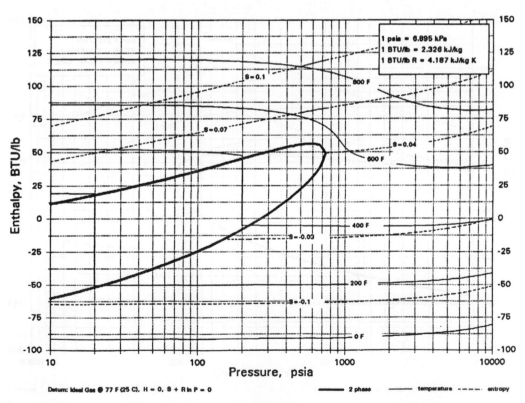

Datum: Ideal Gas @ 77 F (25 C), H = 0, S + R ln P = 0   2 phase   temperature   entropy

82

1. Molecular Weight, lb/mol.......... 58.933

2. Freezing Point, F........................ 2723

3. Boiling Point, F.......................... 4090.7

4. Density @ 20 C, g/cm^3............ 8.9

5. Density @ 68 F, lb/ft^3.............. 555.61

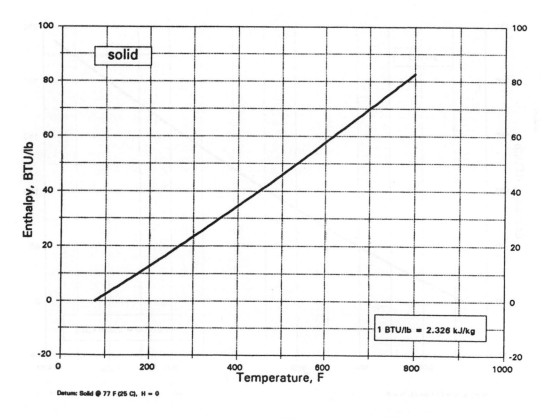

Datum: Solid @ 77 F (25 C), H = 0

1. Molecular Weight, lb/mol.......... 129.839

2. Freezing Point, F........................ 1355

3. Boiling Point, F.......................... 1922

4. Density @ 36 C, g/cm^3............ 3.36

5. Density @ 97 F, lb/ft^3.............. 209.76

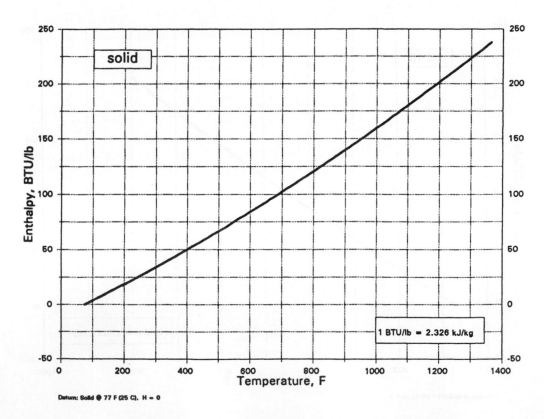

Datum: Solid @ 77 F (25 C), H = 0

1. Molecular Weight, lb/mol.......... 172.971

2. Freezing Point, F........................ 12.2

3. Boiling Point, F.......................... 176

4. Density @ C, g/cm^3............. ---

5. Density @ F, lb/ft^3.............. ---

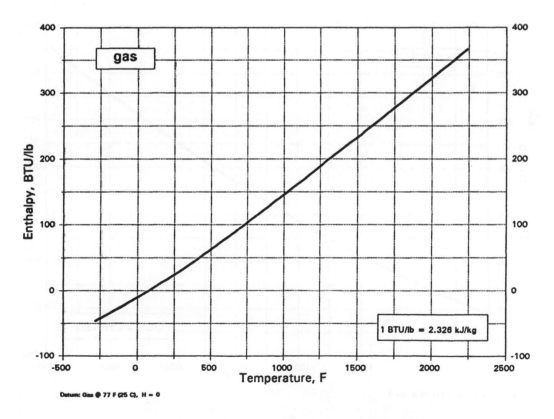

Datum: Gas @ 77 F (25 C), H = 0

85

1. Molecular Weight, lb/mol.......... 51.996

2. Freezing Point, F....................... 3464.6

3. Boiling Point, F......................... 4652.3

4. Density @ 28 C, g/cm^3............ 7.2

5. Density @ 82 F, lb/ft^3.............. 449.48

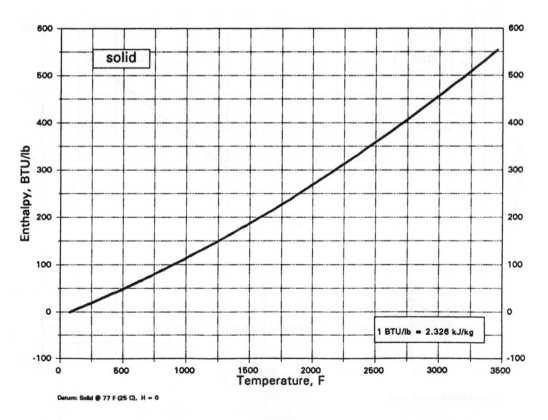

Datum: Solid @ 77 F (25 C), H = 0

86

1. Molecular Weight, lb/mol.......... 220.059

2. Freezing Point, F....................... 302.9

3. Boiling Point, F......................... 303.8

4. Density @ 20 C, g/cm^3............ 1.77

5. Density @ 68 F, lb/ft^3............. 110.5

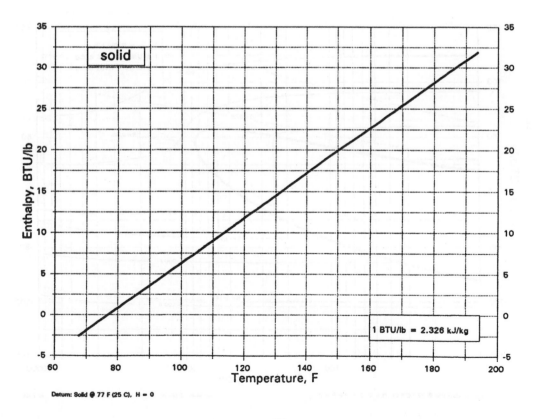

Datum: Solid @ 77 F (25 C), H = 0

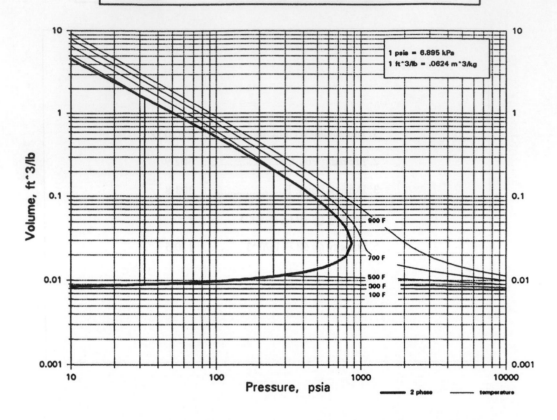

1 psia = 6.895 kPa
1 ft^3/lb = .0624 m^3/kg

900 F
700 F
500 F
300 F
100 F

Volume, ft^3/lb

Pressure, psia

2 phase   temperature

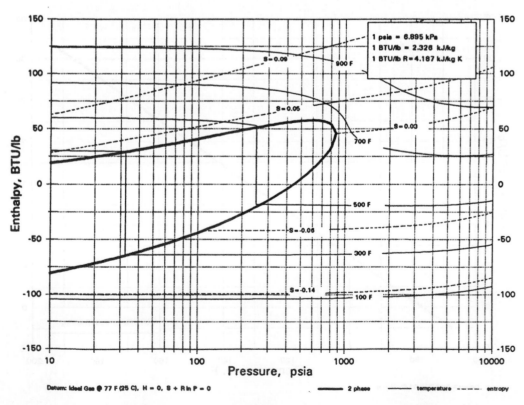

1 psia = 6.895 kPa
1 BTU/lb = 2.326 kJ/kg
1 BTU/lb R = 4.187 kJ/kg K

S = 0.09
900 F
S = 0.05
S = 0.03
700 F
500 F
S = -0.06
300 F
S = -0.14
100 F

Enthalpy, BTU/lb

Pressure, psia

Datum: Ideal Gas @ 77 F (25 C), H = 0, S + R ln P = 0

2 phase   temperature   entropy

1. Molecular Weight, lb/mol.......... 132.905

2. Freezing Point, F....................... 83.3

3. Boiling Point, F.......................... 1274

4. Density @ 15 C, g/cm^3............. 1.88

5. Density @ 59 F, lb/ft^3.............. 117.36

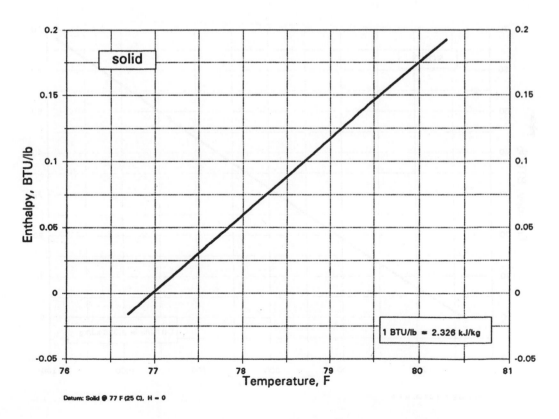

Datum: Solid @ 77 F (25 C), H = 0

1. Molecular Weight, lb/mol.......... 212.809

2. Freezing Point, F....................... 1176.8

3. Boiling Point, F.......................... 2372

4. Density @ 20 C, g/cm^3............. 4.44

5. Density @ 68 F, lb/ft^3.............. 277.18

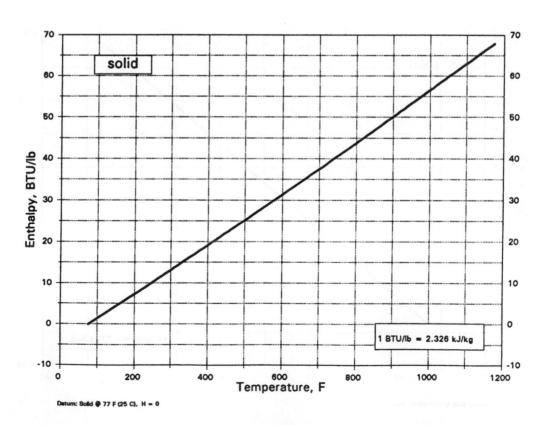

solid

1 BTU/lb = 2.326 kJ/kg

Datum: Solid @ 77 F (25 C), H = 0

1. Molecular Weight, lb/mol.......... 168.358

2. Freezing Point, F........................ 1194.8

3. Boiling Point, F......................... 2372

4. Density @ 20 C, g/cm^3............ 3.99

5. Density @ 68 F, lb/ft^3.............. 249.09

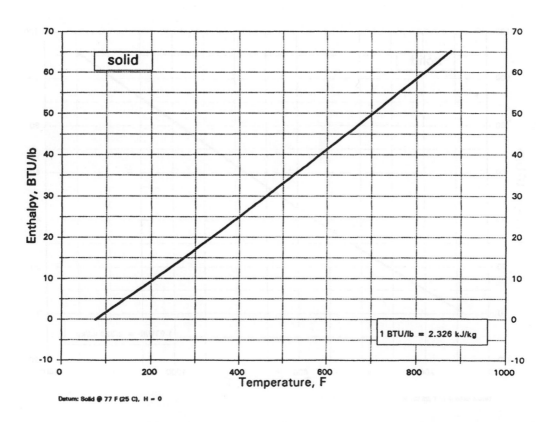

solid

1 BTU/lb = 2.326 kJ/kg

Enthalpy, BTU/lb

Temperature, F

Datum: Solid @ 77 F (25 C), H = 0

1. Molecular Weight, lb/mol.......... 151.904

2. Freezing Point, F....................... 1261.4

3. Boiling Point, F......................... 2283.8

4. Density @ 20 C, g/cm^3............ 4.12

5. Density @ 68 F, lb/ft^3.............. 257.2

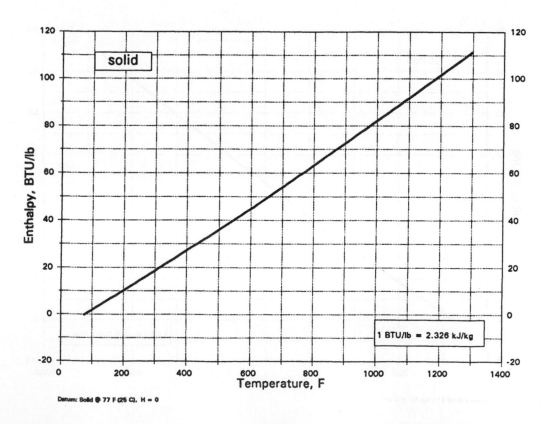

solid

1 BTU/lb = 2.326 kJ/kg

Datum: Solid @ 77 F (25 C), H = 0

1. Molecular Weight, lb/mol.......... 259.81

2. Freezing Point, F...................... 1149.8

3. Boiling Point, F......................... 2336

4. Density @ 25 C, g/cm^3............. 4.51

5. Density @ 77 F, lb/ft^3.............. 281.55

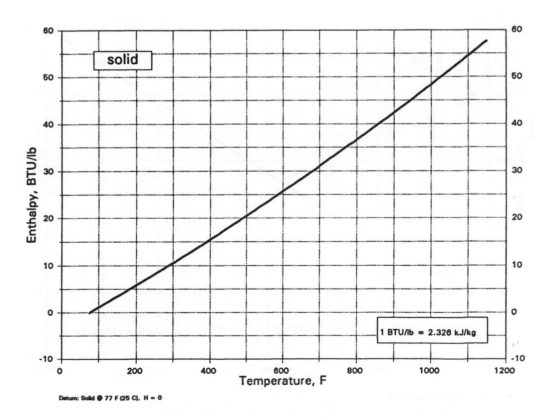

solid

Enthalpy, BTU/lb

Temperature, F

1 BTU/lb = 2.326 kJ/kg

Datum: Solid @ 77 F (25 C), H = 0

1. Molecular Weight, lb/mol.......... 63.546

2. Freezing Point, F....................... 1984.3

3. Boiling Point, F......................... 5210.3

4. Density @ 20 C, g/cm^3............. 8.92

5. Density @ 68 F, lb/ft^3.............. 556.86

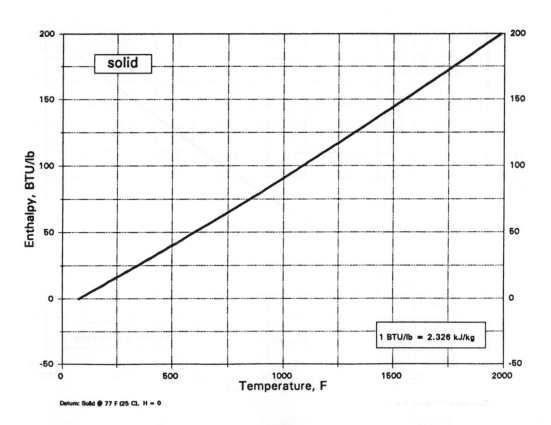

Datum: Solid @ 77 F (25 C), H = 0

1. Molecular Weight, lb/mol.......... 143.45

2. Freezing Point, F....................... 939.2

3. Boiling Point, F......................... 2471

4. Density @ 20 C, g/cm^3............ 4.98

5. Density @ 68 F, lb/ft^3.............. 310.89

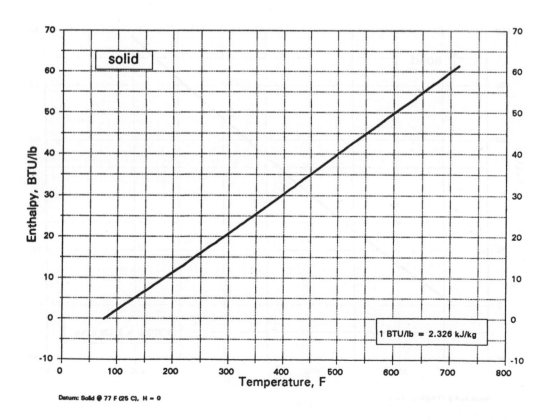

Datum: Solid @ 77 F (25 C), H = 0

1. Molecular Weight, lb/mol........... 98.999

2. Freezing Point, F....................... 805.7

3. Boiling Point, F......................... 2714

4. Density @ 20 C, g/cm^3............ 4.14

5. Density @ 68 F, lb/ft^3.............. 258.45

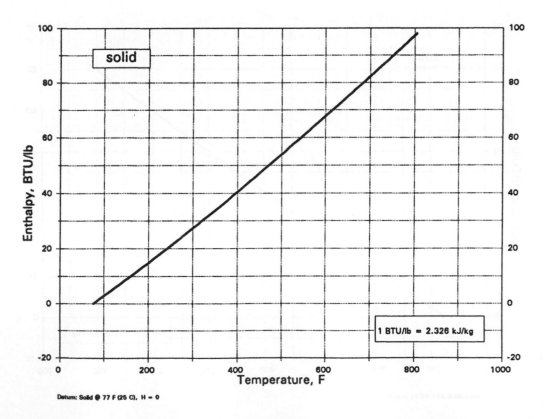

Enthalpy, BTU/lb vs Temperature, F

solid

1 BTU/lb = 2.326 kJ/kg

Datum: Solid @ 77 F (25 C), H = 0

1. Molecular Weight, lb/mol.......... 134.451

2. Freezing Point, F....................... 1171.4

3. Boiling Point, F......................... 1819.4

4. Density @ 25 C, g/cm^3............. 3.39

5. Density @ 77 F, lb/ft^3.............. 211.63

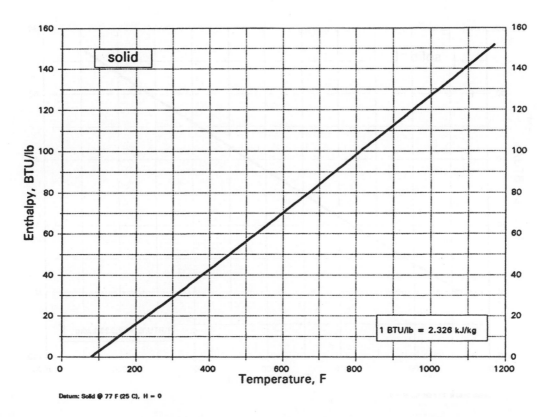

Datum: Solid @ 77 F (25 C), H = 0

1. Molecular Weight, lb/mol.......... 190.45

2. Freezing Point, F........................ 1121

3. Boiling Point, F.......................... 2436.8

4. Density @ 20 C, g/cm^3............. 5.62

5. Density @ 68 F, lb/ft^3.............. 350.84

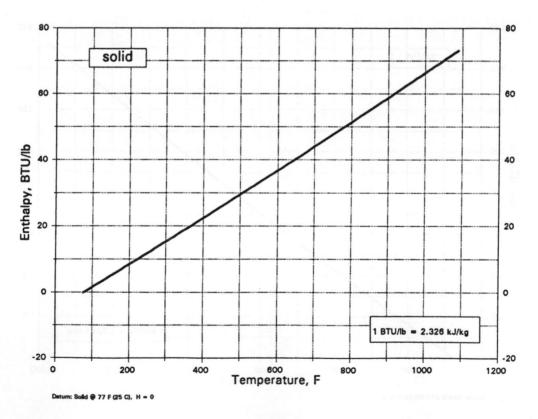

Datum: Solid @ 77 F (25 C), H = 0

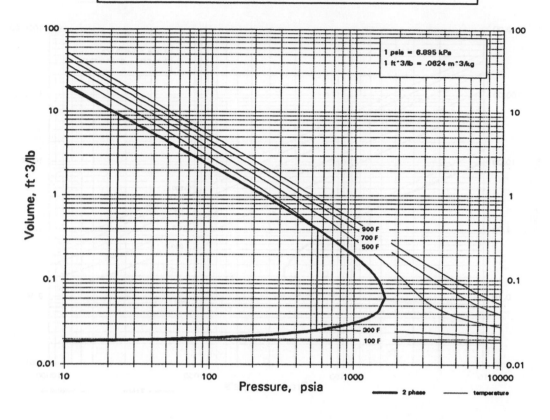

1 psia = 6.895 kPa
1 ft^3/lb = .0624 m^3/kg

900 F
700 F
500 F
300 F
100 F

Volume, ft^3/lb

Pressure, psia

2 phase   temperature

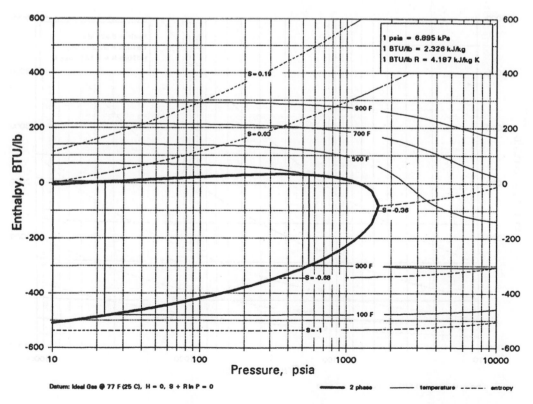

1 psia = 6.895 kPa
1 BTU/lb = 2.326 kJ/kg
1 BTU/lb R = 4.187 kJ/kg K

S = 0.19
S = 0.03
900 F
700 F
500 F
S = -0.36
300 F
S = -0.68
100 F
S = -1

Enthalpy, BTU/lb

Pressure, psia

Datum: Ideal Gas @ 77 F (25 C), H = 0, S + R ln P = 0

2 phase   temperature ---- entropy

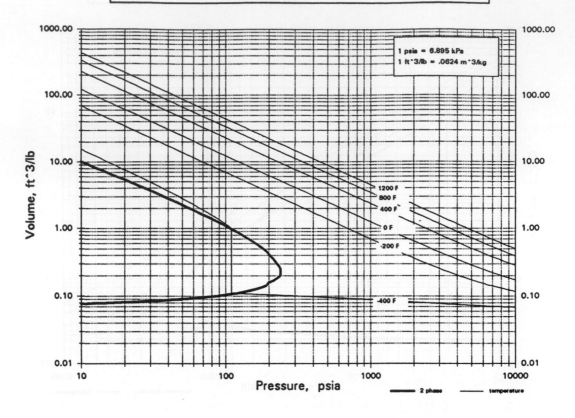

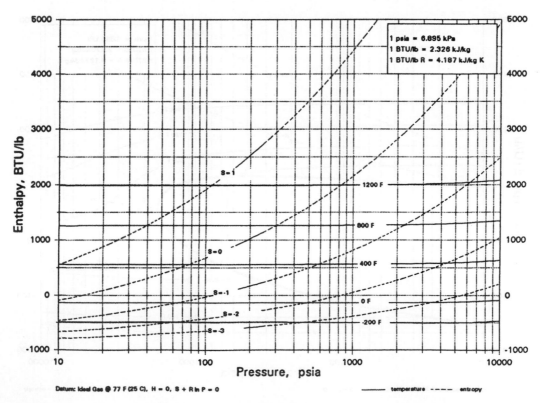

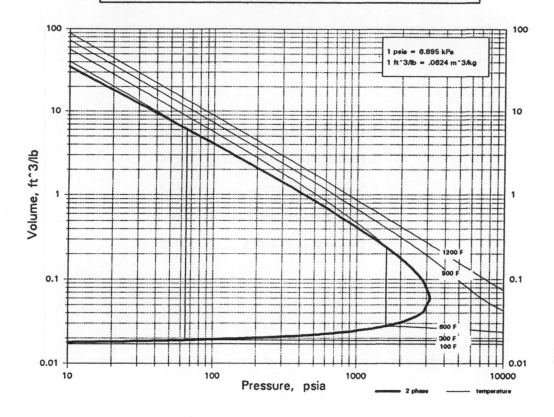

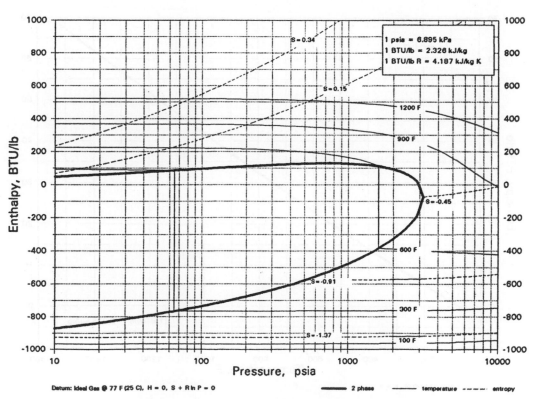

1. Molecular Weight, lb/mol.......... 151.965

2. Freezing Point, F........................ 1511.6

3. Boiling Point, F.......................... 2675.9

4. Density @ 20 C, g/cm^3............ 5.24

5. Density @ 68 F, lb/ft^3.............. 327.12

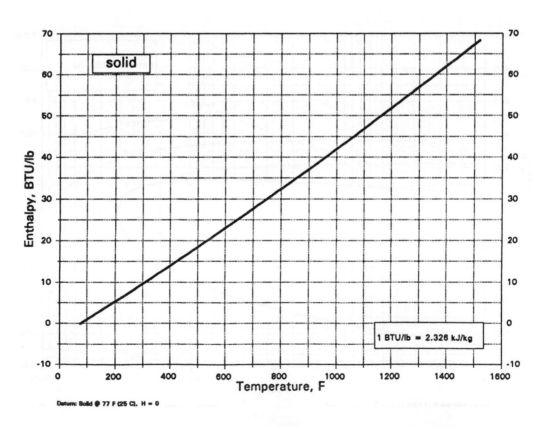

Datum: Solid @ 77 F (25 C), H = 0

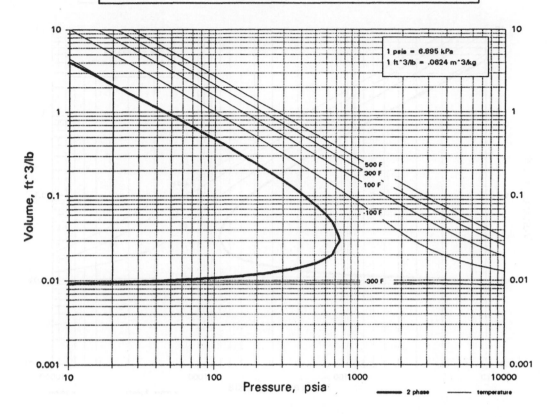

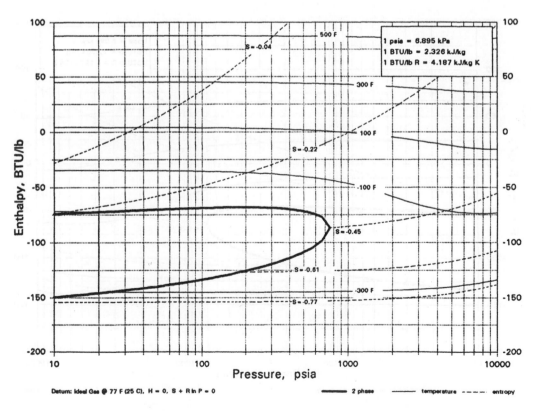

Datum: Ideal Gas @ 77 F (25 C), H = 0, S + R ln P = 0

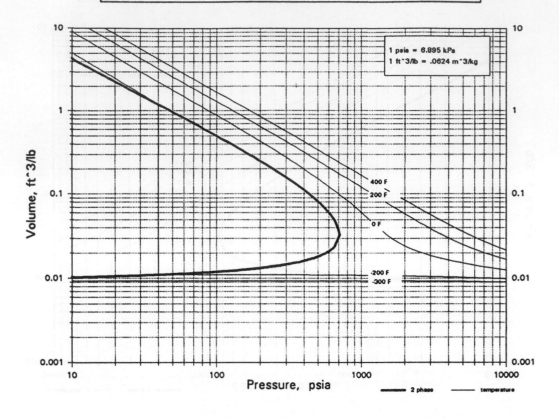

1 psia = 6.895 kPa
1 ft^3/lb = .0624 m^3/kg

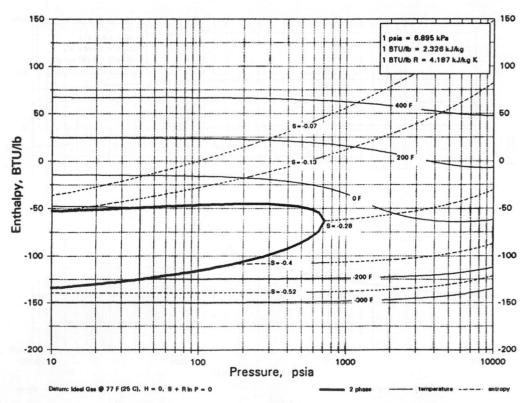

1 psia = 6.895 kPa
1 BTU/lb = 2.326 kJ/kg
1 BTU/lb R = 4.187 kJ/kg K

Datum: Ideal Gas @ 77 F (25 C), H = 0, S + R ln P = 0

1. Molecular Weight, lb/mol.......... 55.847

2. Freezing Point, F....................... 2795

3. Boiling Point, F......................... 4940.3

4. Density @ 20 C, g/cm^3............ 7.86

5. Density @ 68 F, lb/ft^3.............. 490.68

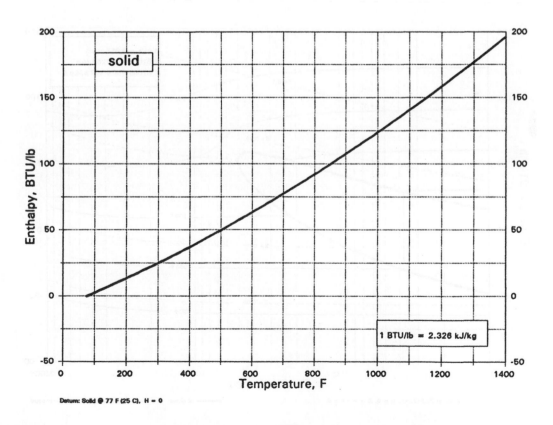

solid

1 BTU/lb = 2.326 kJ/kg

Datum: Solid @ 77 F (25 C), H = 0

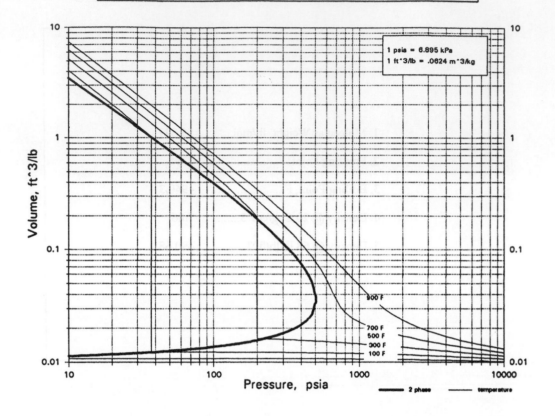

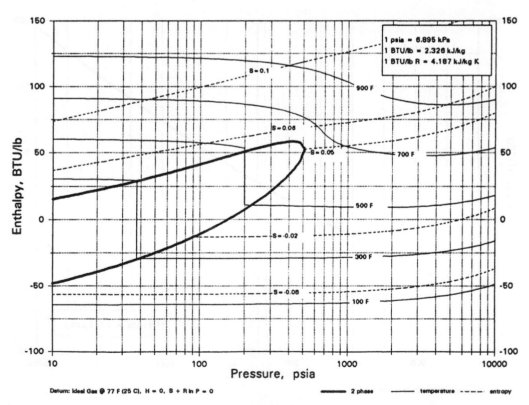

1. Molecular Weight, lb/mol.......... 126.752

2. Freezing Point, F........................ 1241.6

3. Boiling Point, F.......................... 1878.8

4. Density @ 25 C, g/cm^3............. 3.16

5. Density @ 77 F, lb/ft^3.............. 197.27

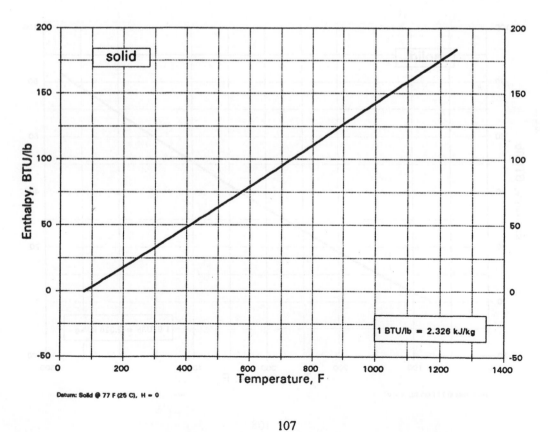

solid

1 BTU/lb = 2.326 kJ/kg

Datum: Solid @ 77 F (25 C), H = 0

1. Molecular Weight, lb/mol.......... 162.205

2. Freezing Point, F........................ 579.2

3. Boiling Point, F.......................... 606.2

4. Density @ 25 C, g/cm^3............ 2.9

5. Density @ 77 F, lb/ft^3.............. 181.04

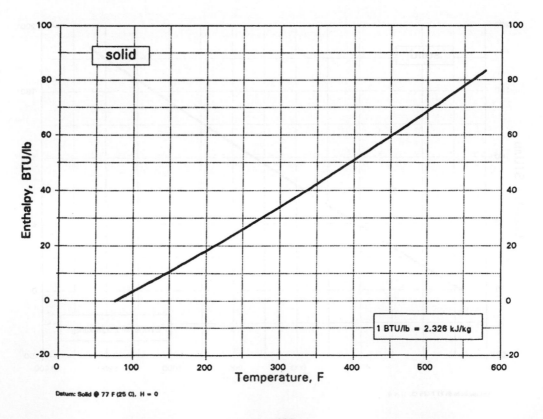

1 BTU/lb = 2.326 kJ/kg

Datum: Solid @ 77 F (25 C), H = 0

1. Molecular Weight, lb/mol.......... 223

2. Freezing Point, F....................... 80.6

3. Boiling Point, F......................... 1122.5

4. Density @ C, g/cm^3............ ---

5. Density @ F, lb/ft^3.............. ---

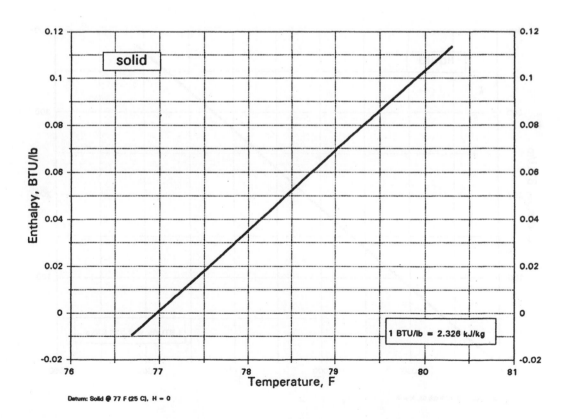

Datum: Solid @ 77 F (25 C). H = 0

1. Molecular Weight, lb/mol.......... 69.723

2. Freezing Point, F........................ 85.5

3. Boiling Point, F......................... 4070.9

4. Density @ 30 C, g/cm^3............ 6.1

5. Density @ 86 F, lb/ft^3.............. 380.81

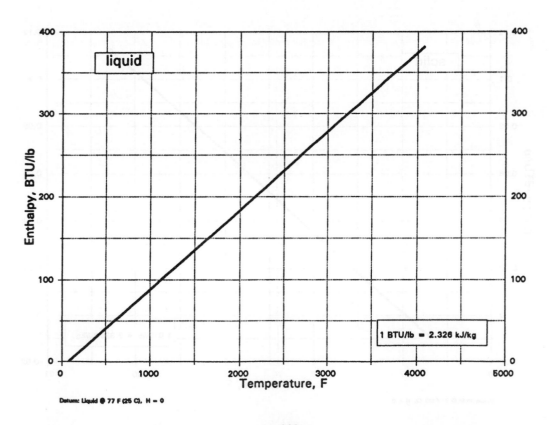

liquid

1 BTU/lb = 2.326 kJ/kg

Enthalpy, BTU/lb

Temperature, F

Datum: Liquid @ 77 F (25 C), H = 0

1. Molecular Weight, lb/mol.......... 176.081

2. Freezing Point, F........................ 171.9

3. Boiling Point, F......................... 393.8

4. Density @ 26 C, g/cm^3............ 2.47

5. Density @ 79 F, lb/ft^3.............. 154.2

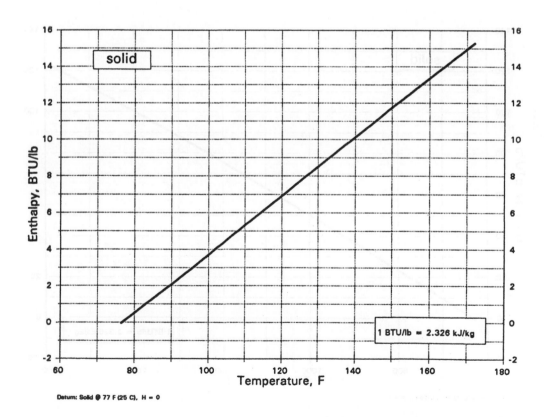

Datum: Solid @ 77 F (25 C), H = 0

1. Molecular Weight, lb/mol.......... 157.25

2. Freezing Point, F....................... 2397.2

3. Boiling Point, F......................... 2726.3

4. Density @ 20 C, g/cm^3............ 7.9

5. Density @ 68 F, lb/ft^3.............. 493.18

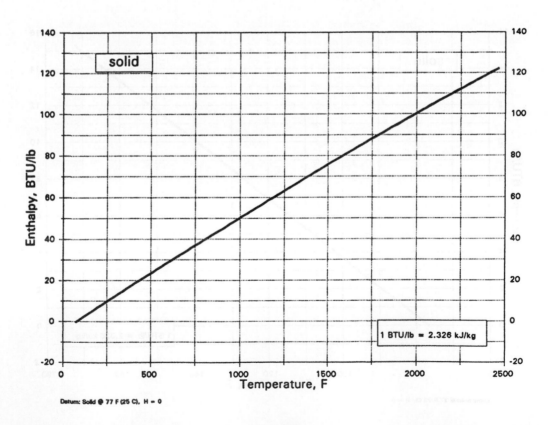

Datum: Solid @ 77 F (25 C), H = 0

1. Molecular Weight, lb/mol.......... 72.61

2. Freezing Point, F...................... 1720.8

3. Boiling Point, F......................... 5165.3

4. Density @ 20 C, g/cm^3............ 5.35

5. Density @ 68 F, lb/ft^3.............. 333.99

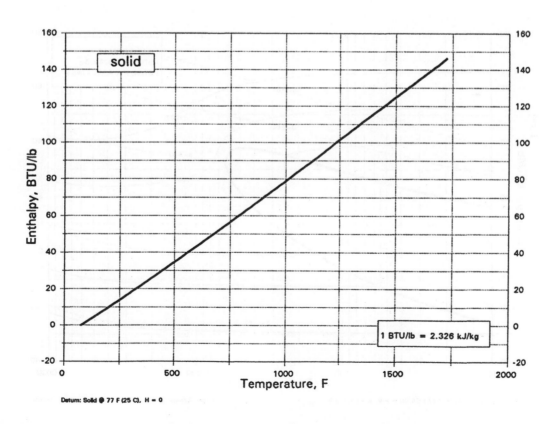

solid

1 BTU/lb = 2.326 kJ/kg

Datum: Solid @ 77 F (25 C), H = 0

113

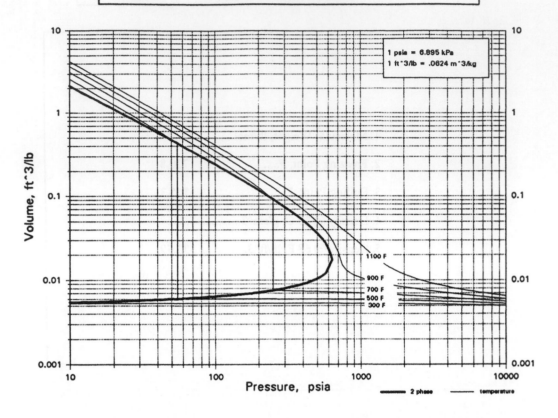

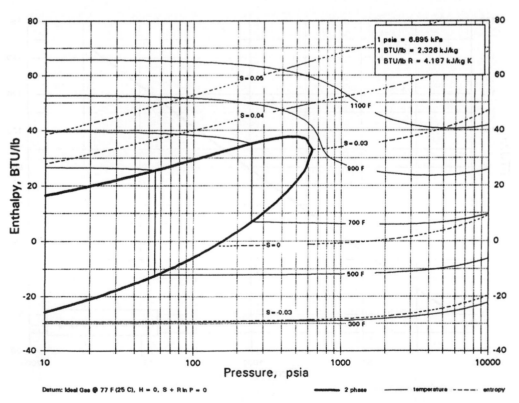

Datum: Ideal Gas @ 77 F (25 C), H = 0, S + R ln P = 0

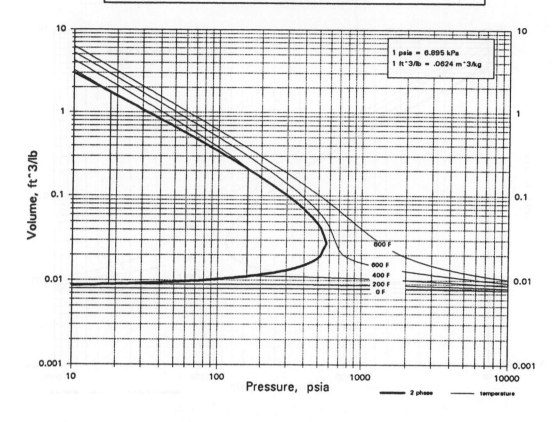

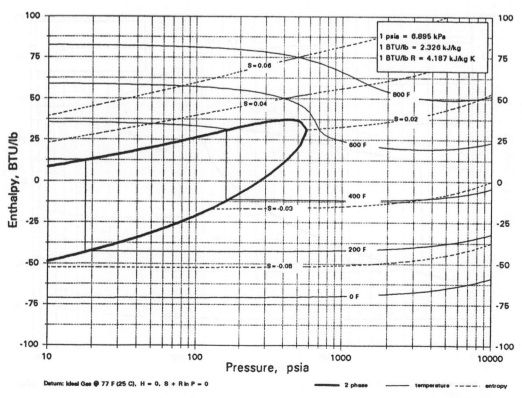

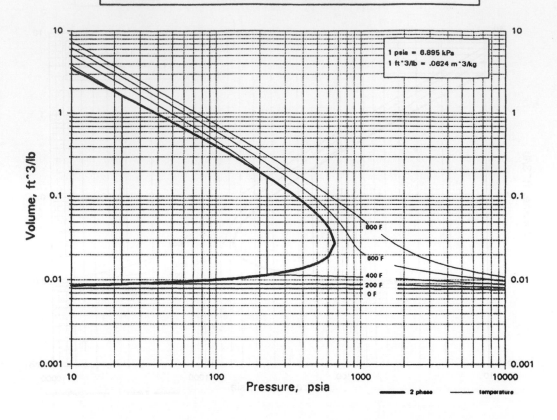

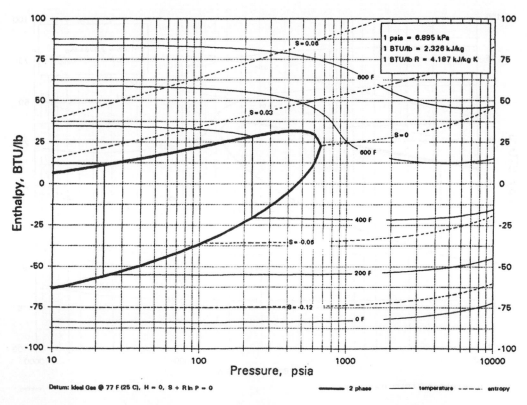

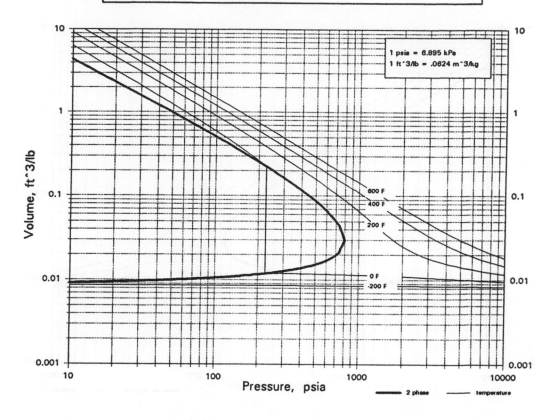

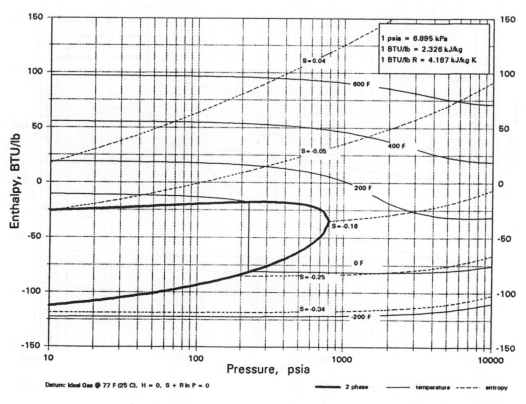

Datum: Ideal Gas @ 77 F (25 C), H = 0, S + R ln P = 0

117

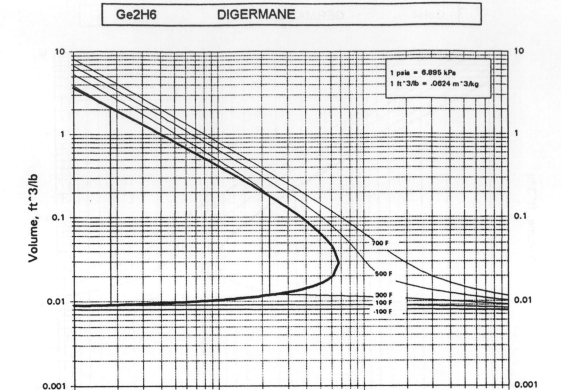

1 psia = 6.895 kPa
1 ft^3/lb = .0624 m^3/kg

700 F
500 F
300 F
100 F
-100 F

Volume, ft^3/lb

Pressure, psia

2 phase   temperature

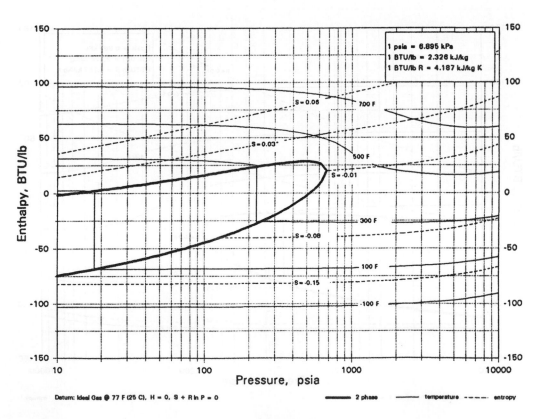

1 psia = 6.895 kPa
1 BTU/lb = 2.326 kJ/kg
1 BTU/lb R = 4.187 kJ/kg K

S = 0.06   700 F
S = 0.03
500 F
S = -0.01
300 F
S = -0.08
100 F
S = -0.15
-100 F

Enthalpy, BTU/lb

Pressure, psia

Datum: Ideal Gas @ 77 F (25 C), H = 0, S + R ln P = 0

2 phase   temperature   entropy

118

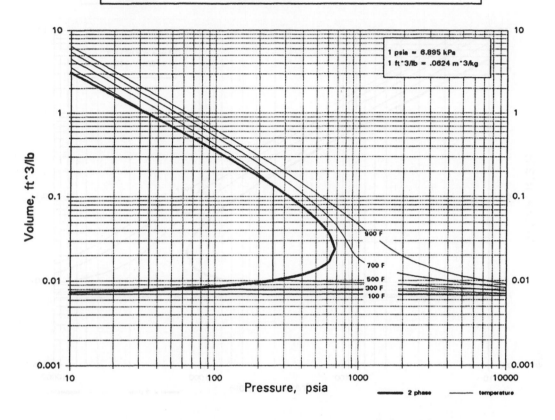

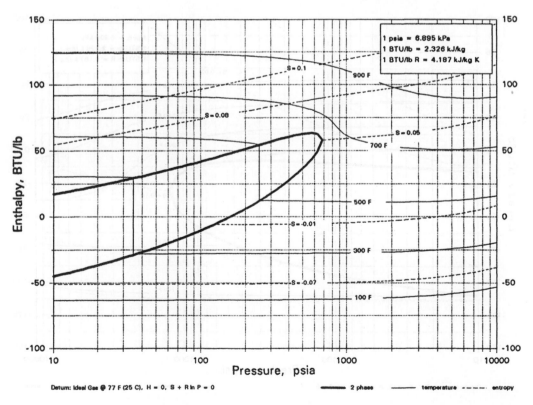

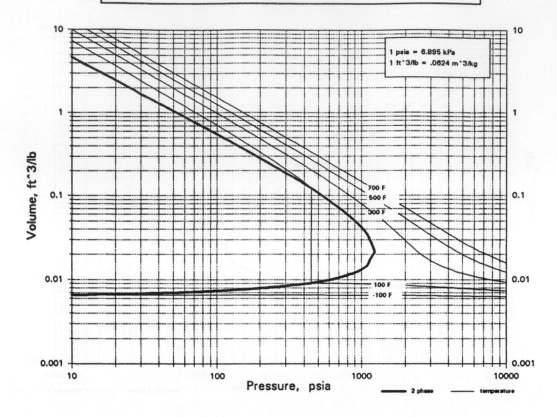

1 psia = 6.895 kPa
1 ft^3/lb = .0624 m^3/kg

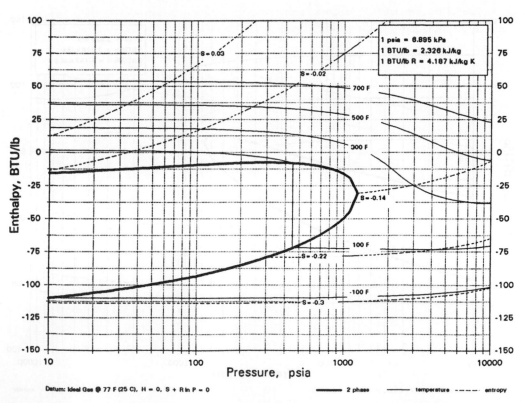

1 psia = 6.895 kPa
1 BTU/lb = 2.326 kJ/kg
1 BTU/lb R = 4.187 kJ/kg K

Datum: Ideal Gas @ 77 F (25 C), H = 0, S + R ln P = 0

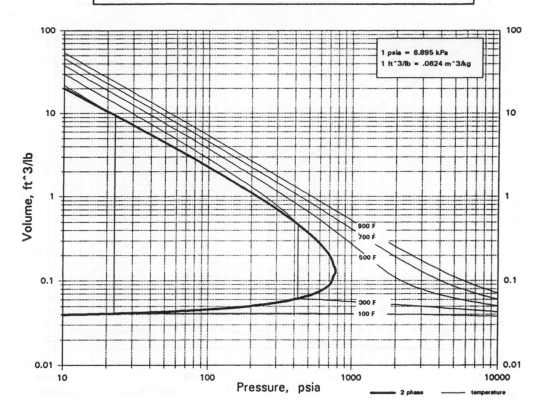

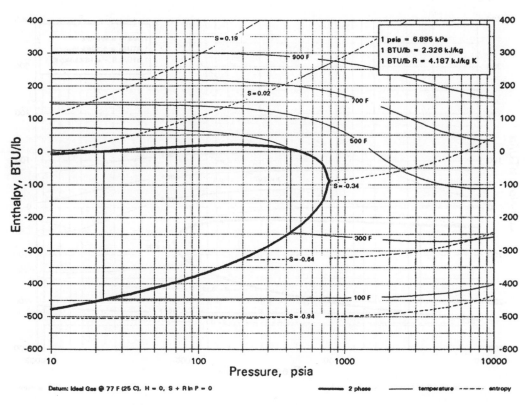

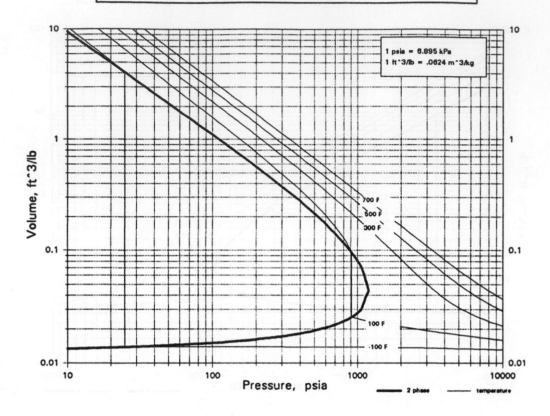

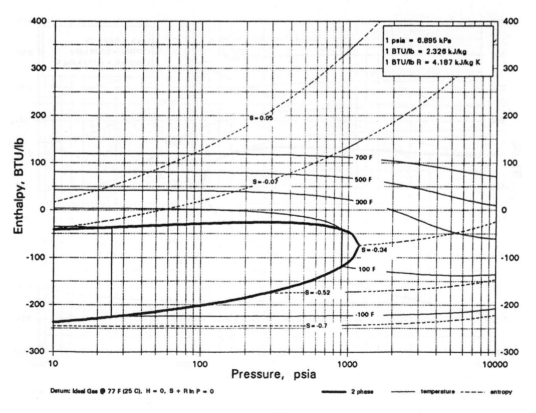

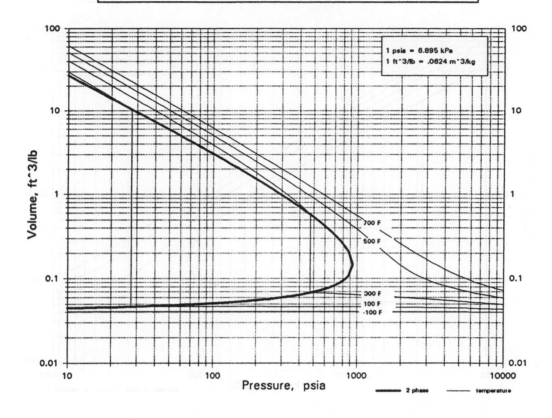

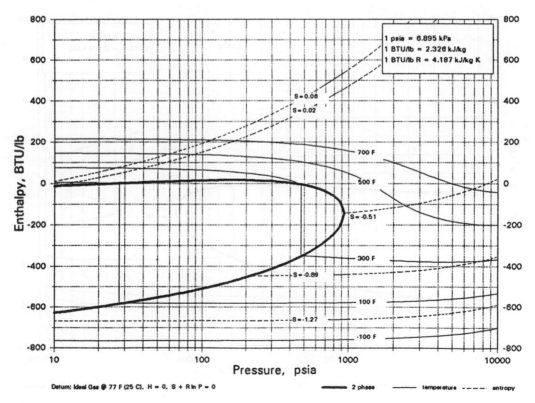

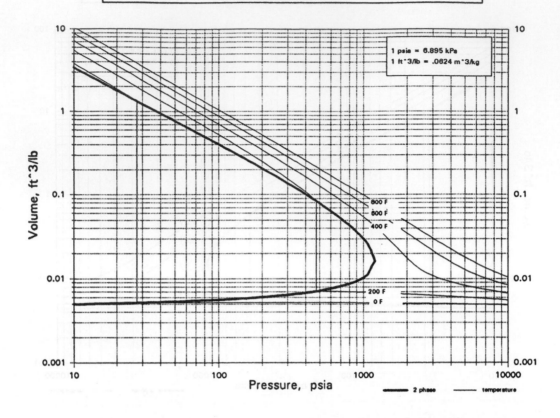

1 psia = 6.895 kPa
1 ft^3/lb = .0624 m^3/kg

800 F
600 F
400 F

200 F
0 F

Volume, ft^3/lb

Pressure, psia

2 phase   temperature

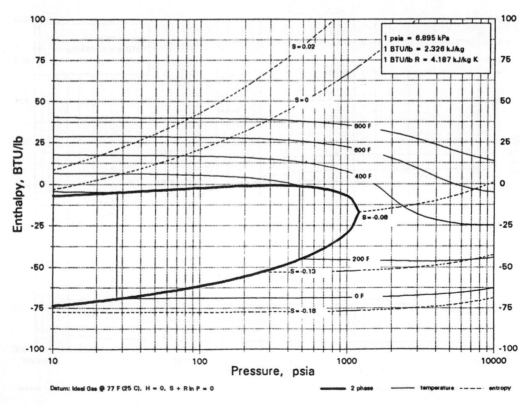

1 psia = 6.895 kPa
1 BTU/lb = 2.326 kJ/kg
1 BTU/lb R = 4.187 kJ/kg K

S = 0.02

S = 0

800 F
600 F
400 F

S = -0.08

200 F
S = -0.13

0 F

S = -0.18

Enthalpy, BTU/lb

Pressure, psia

Datum: Ideal Gas @ 77 F (25 C), H = 0, S + R ln P = 0

2 phase   temperature   entropy

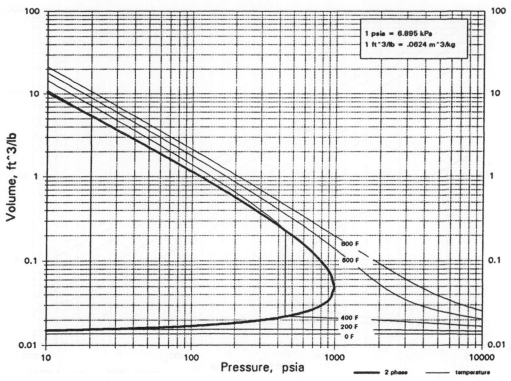

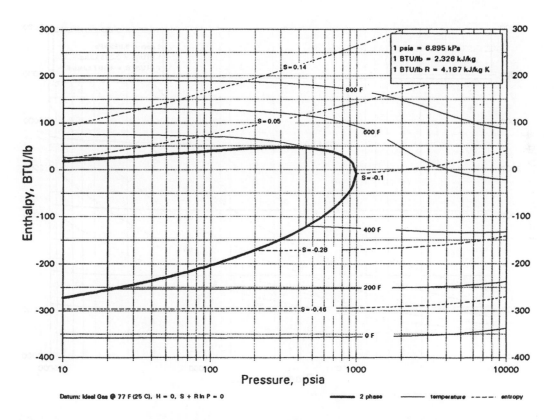

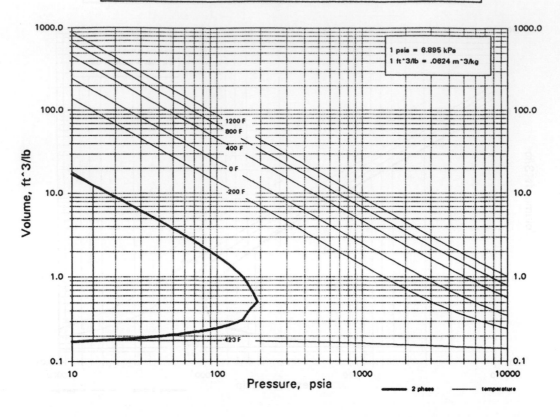

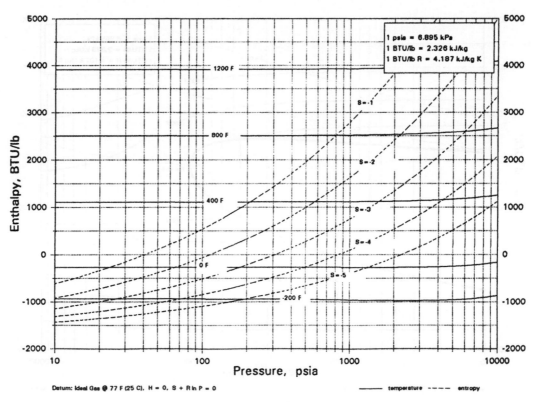

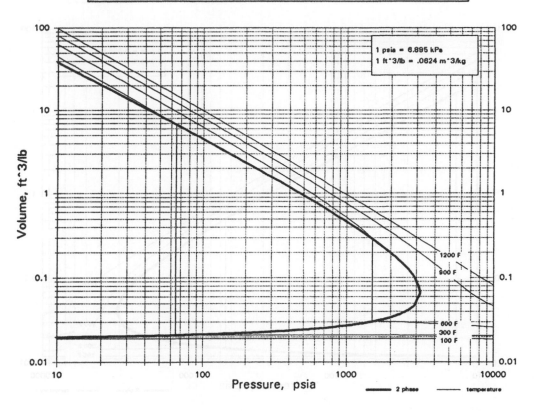

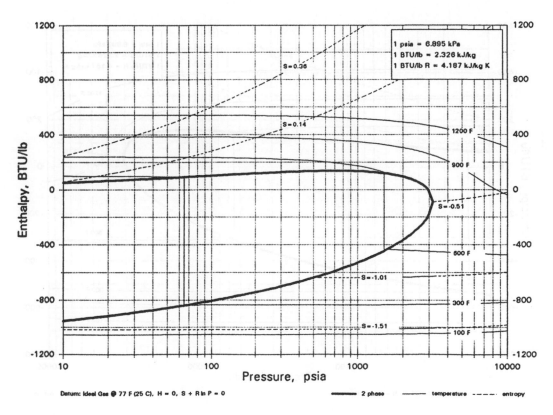

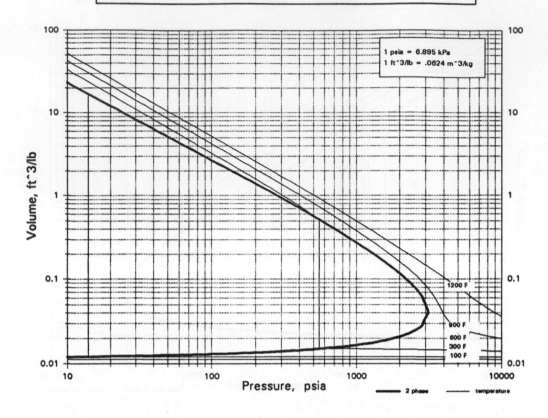

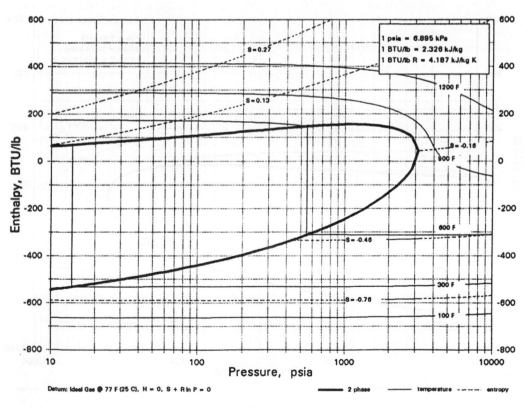

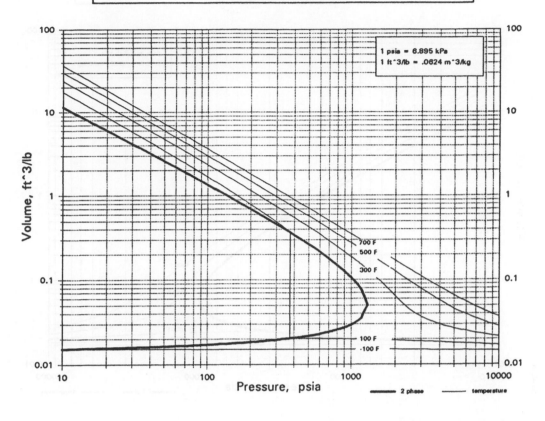

1 psia = 6.895 kPa
1 ft^3/lb = .0624 m^3/kg

700 F
500 F
300 F

100 F
-100 F

2 phase   temperature

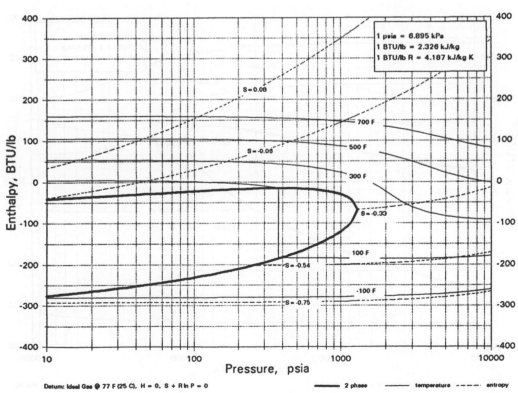

1 psia = 6.895 kPa
1 BTU/lb = 2.326 kJ/kg
1 BTU/lb R = 4.187 kJ/kg K

S = 0.08

S = -0.06

700 F

500 F

300 F

S = -0.33

100 F

S = -0.54

-100 F

S = -0.75

Datum: Ideal Gas @ 77 F (25 C). H = 0, S + R ln P = 0   2 phase   temperature ----- entropy

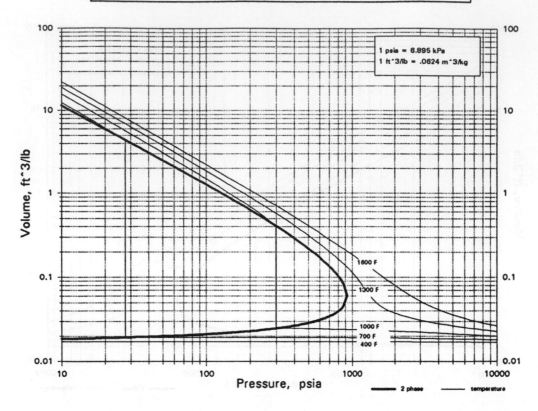

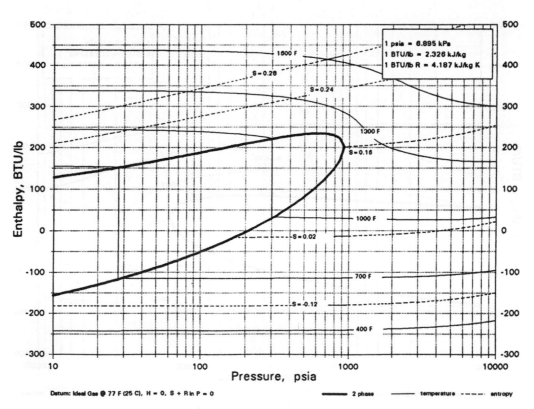

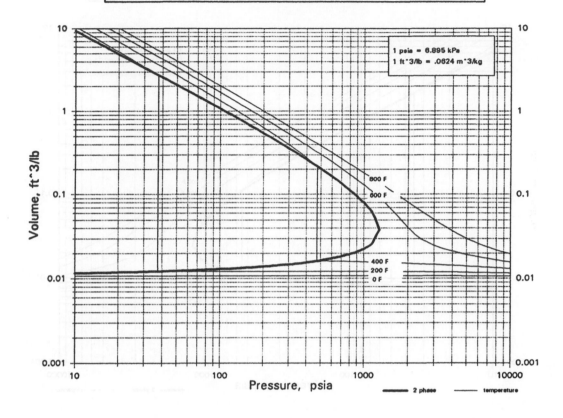

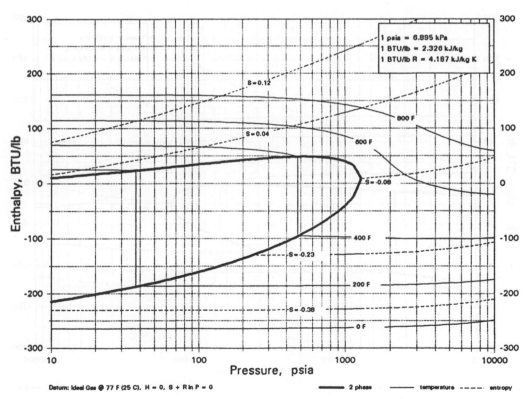

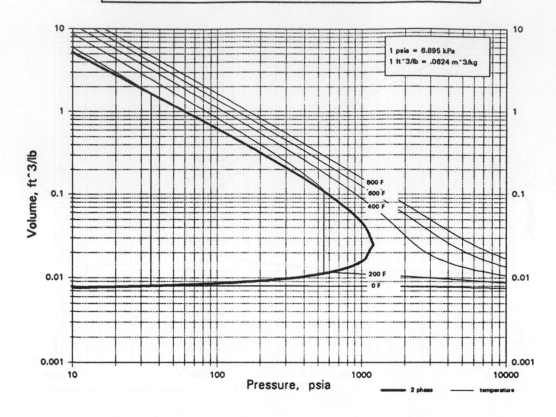

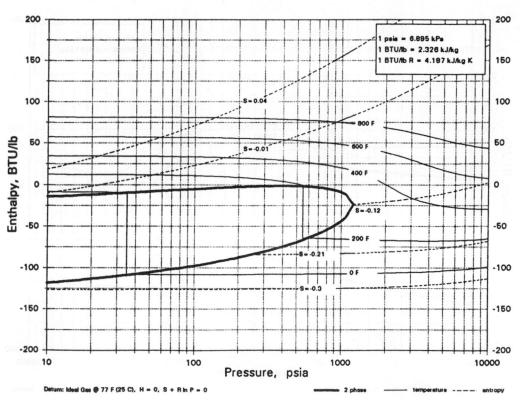

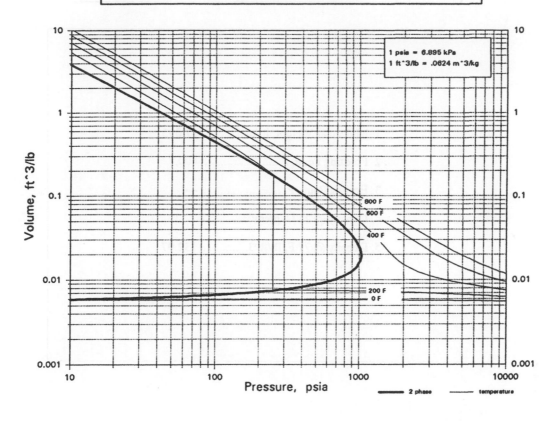

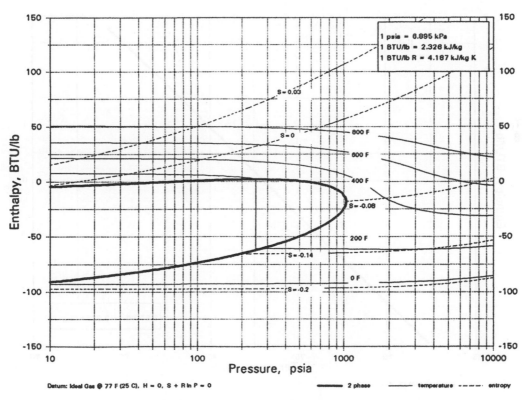

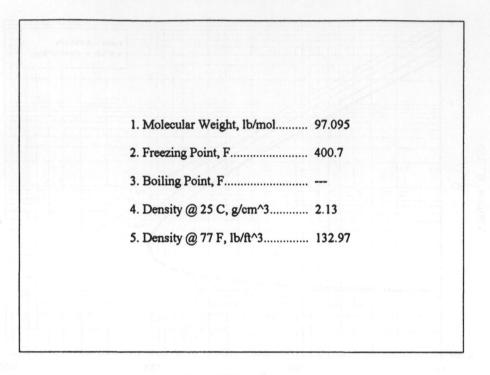

1. Molecular Weight, lb/mol.......... 97.095

2. Freezing Point, F....................... 400.7

3. Boiling Point, F.......................... ---

4. Density @ 25 C, g/cm^3............ 2.13

5. Density @ 77 F, lb/ft^3.............. 132.97

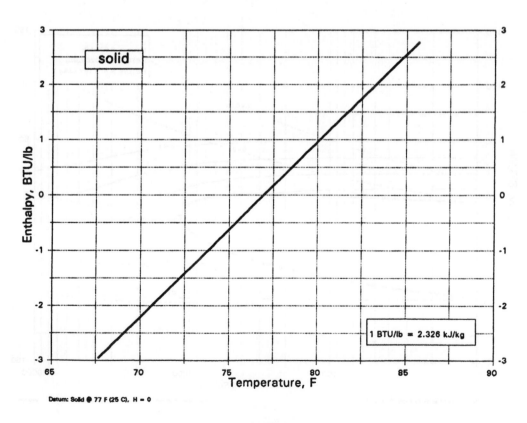

Datum: Solid @ 77 F (25 C), H = 0

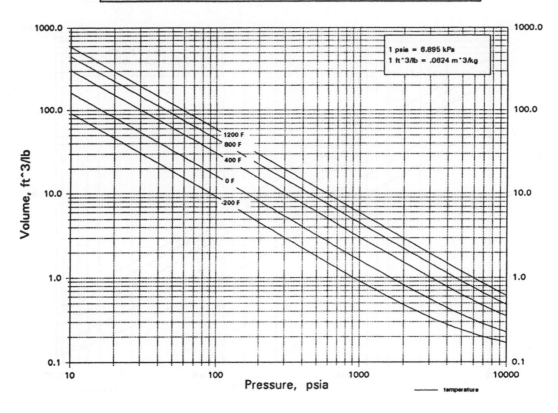

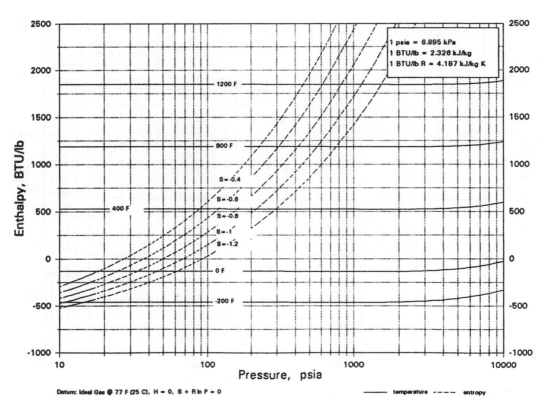

He   HELIUM-4

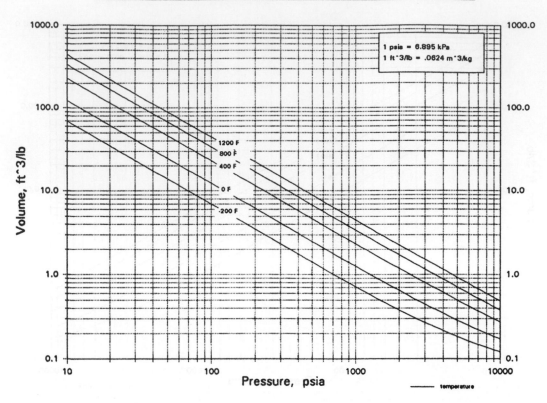

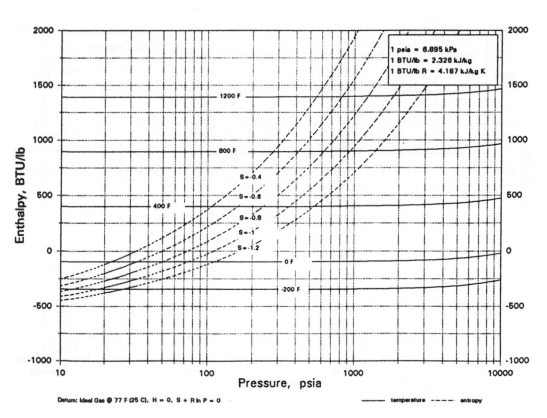

Datum: Ideal Gas @ 77 F (25 C), H = 0, S + R ln P = 0   —— temperature - - - - entropy

136

1. Molecular Weight, lb/mol.......... 178.49

2. Freezing Point, F....................... 4051.4

3. Boiling Point, F......................... 10268.3

4. Density @ 20 C, g/cm^3............ 13.31

5. Density @ 68 F, lb/ft^3.............. 830.91

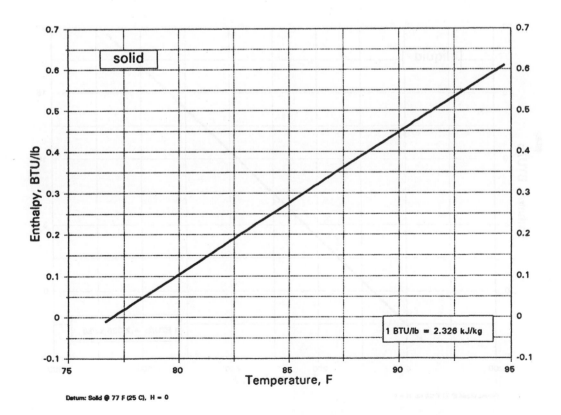

Datum: Solid @ 77 F (25 C), H = 0

1. Molecular Weight, lb/mol.......... 200.59

2. Freezing Point, F........................ -38

3. Boiling Point, F........................... 673.8

4. Density @ 20 C, g/cm^3............. 13.55

5. Density @ 68 F, lb/ft^3.............. 845.9

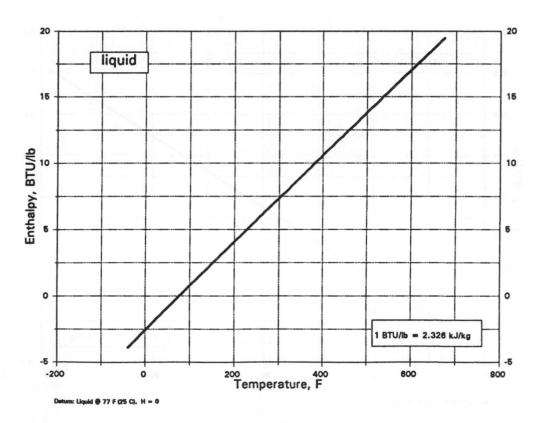

Datum: Liquid @ 77 F (25 C), H = 0

1. Molecular Weight, lb/mol.......... 360.398

2. Freezing Point, F........................ 458.6

3. Boiling Point, F......................... 606.2

4. Density @ 25 C, g/cm^3............. 6.11

5. Density @ 77 F, lb/ft^3.............. 381.43

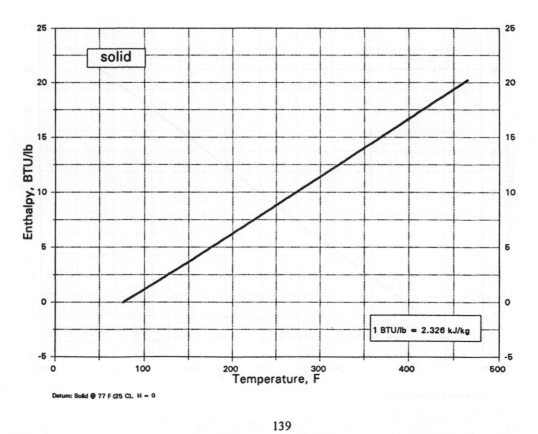

Datum: Solid @ 77 F (25 C), H = 0

1. Molecular Weight, lb/mol.......... 271.495

2. Freezing Point, F........................ 530.6

3. Boiling Point, F.......................... 579.2

4. Density @ 25 C, g/cm^3............ 5.44

5. Density @ 77 F, lb/ft^3.............. 339.61

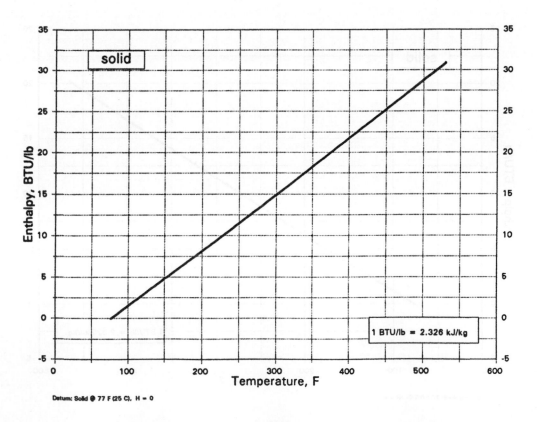

Datum: Solid @ 77 F (25 C). H = 0

1. Molecular Weight, lb/mol.......... 454.399

2. Freezing Point, F....................... 498.2

3. Boiling Point, F.......................... 669.2

4. Density @ 127 C, g/cm^3.......... 6.09

5. Density @ 261 F, lb/ft^3............. 380.19

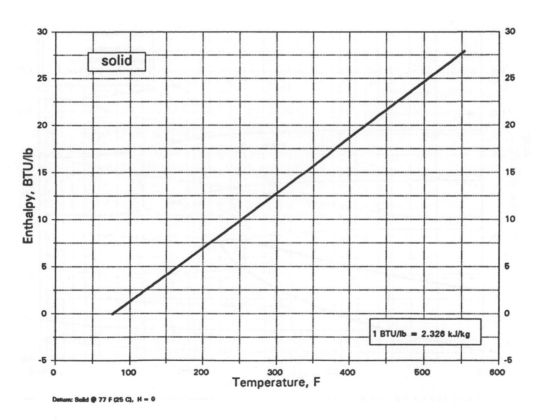

solid

1 BTU/lb = 2.326 kJ/kg

Datum: Solid @ 77 F (25 C), H = 0

141

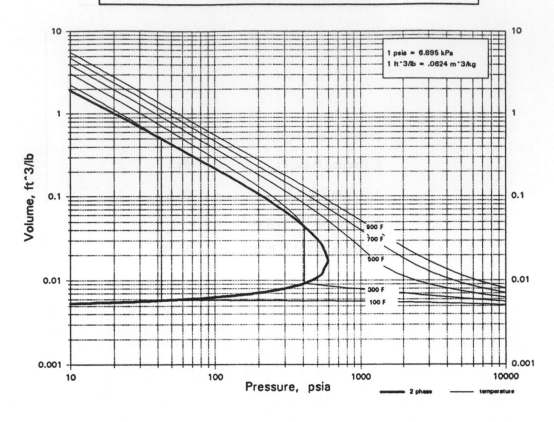

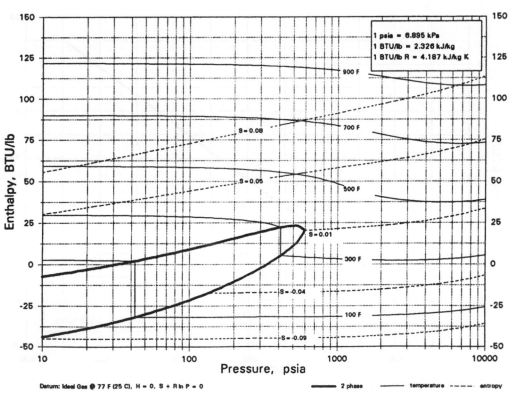

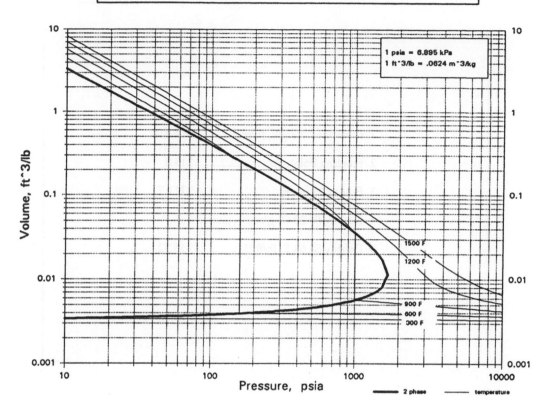

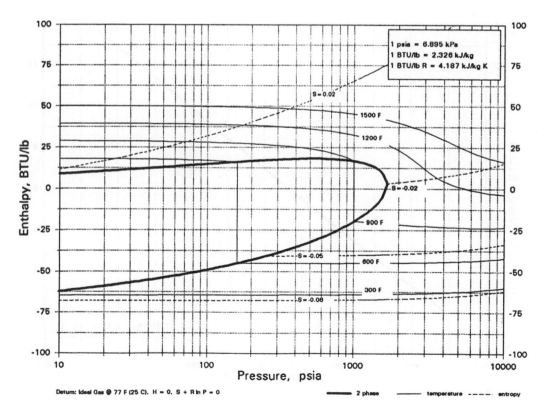

1. Molecular Weight, lb/mol.......... 114.818

2. Freezing Point, F....................... 313.8

3. Boiling Point, F.......................... 3721.7

4. Density @ 20 C, g/cm^3............. 7.3

5. Density @ 68 F, lb/ft^3.............. 455.72

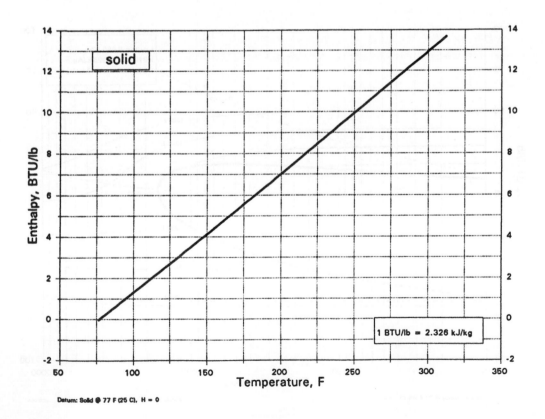

Datum: Solid @ 77 F (25 C), H = 0

144

1. Molecular Weight, lb/mol.......... 192.22

2. Freezing Point, F....................... 4434.8

3. Boiling Point, F......................... 7550.3

4. Density @ 20 C, g/cm^3............ 22.42

5. Density @ 68 F, lb/ft^3.............. 1399.63

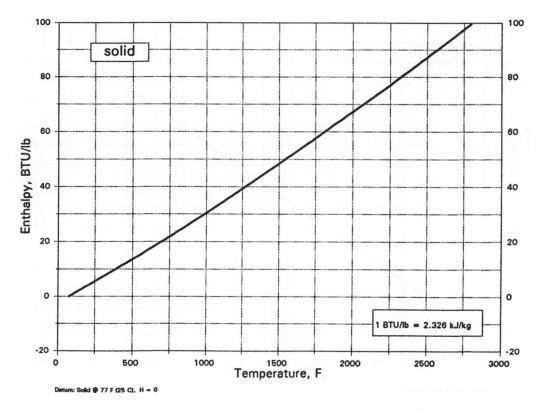

solid

1 BTU/lb = 2.326 kJ/kg

Enthalpy, BTU/lb

Temperature, F

Datum: Solid @ 77 F (25 C), H = 0

1. Molecular Weight, lb/mol.......... 39.098

2. Freezing Point, F...................... 145.7

3. Boiling Point, F......................... 1406.9

4. Density @ 20 C, g/cm^3............ 0.86

5. Density @ 68 F, lb/ft^3.............. 53.69

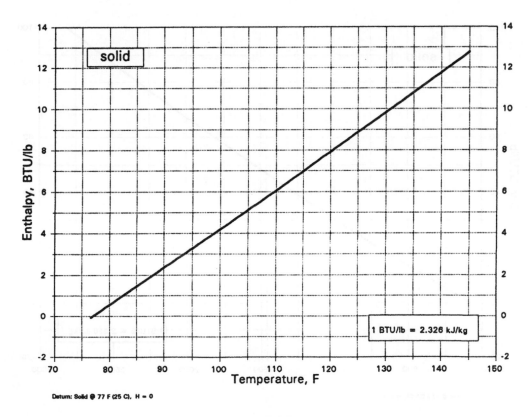

1 BTU/lb = 2.326 kJ/kg

Datum: Solid @ 77 F (25 C), H = 0

1. Molecular Weight, lb/mol.......... 119.002

2. Freezing Point, F........................ 1346

3. Boiling Point, F......................... 2521.4

4. Density @ 25 C, g/cm^3............ 2.75

5. Density @ 77 F, lb/ft^3.............. 171.68

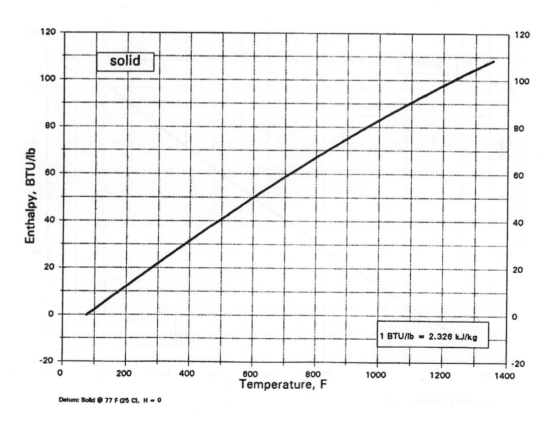

solid

1 BTU/lb = 2.326 kJ/kg

Datum: Solid @ 77 F (25 C), H = 0

1. Molecular Weight, lb/mol.......... 74.551

2. Freezing Point, F....................... 1419.5

3. Boiling Point, F........................ 2580.3

4. Density @ 20 C, g/cm^3............ 1.98

5. Density @ 68 F, lb/ft^3.............. 123.61

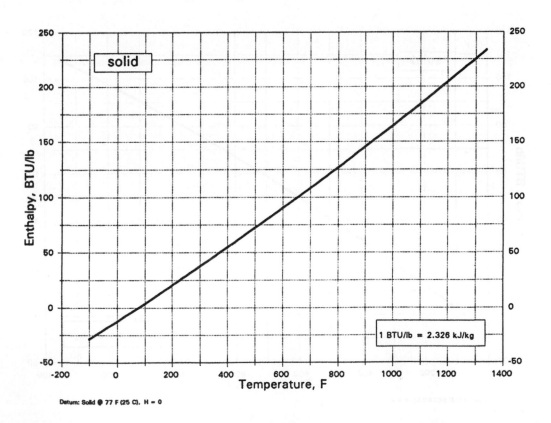

Datum: Solid @ 77 F (25 C), H = 0

1. Molecular Weight, lb/mol.......... 58.097

2. Freezing Point, F........................ 1616

3. Boiling Point, F.......................... 2735.6

4. Density @ 20 C, g/cm^3............ 2.48

5. Density @ 68 F, lb/ft^3.............. 154.82

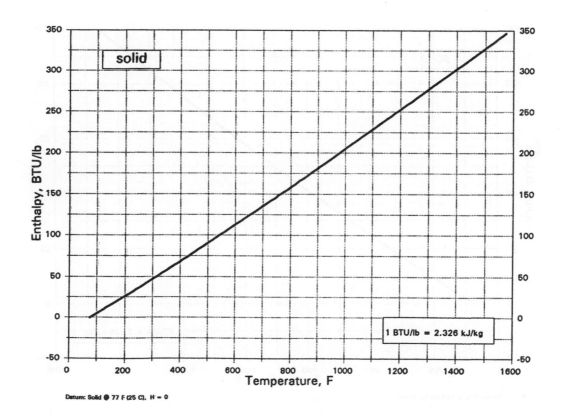

Datum: Solid @ 77 F (25 C). H = 0

149

1. Molecular Weight, lb/mol.......... 166.003

2. Freezing Point, F....................... 1333.4

3. Boiling Point, F.......................... 2415.2

4. Density @ 20 C, g/cm^3............ 3.13

5. Density @ 68 F, lb/ft^3.............. 195.4

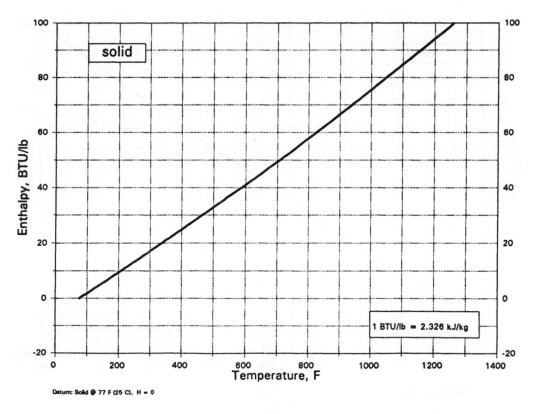

solid

1 BTU/lb = 2.326 kJ/kg

Datum: Solid @ 77 F (25 C), H = 0

1. Molecular Weight, lb/mol.......... 56.106

2. Freezing Point, F...................... 762.5

3. Boiling Point, F........................ 2420.3

4. Density @ 20 C, g/cm^3............ 2.04

5. Density @ 68 F, lb/ft^3.............. 127.35

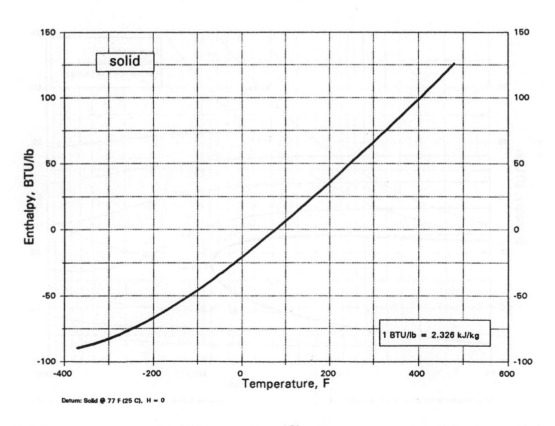

solid

1 BTU/lb = 2.326 kJ/kg

Datum: Solid @ 77 F (25 C), H = 0

151

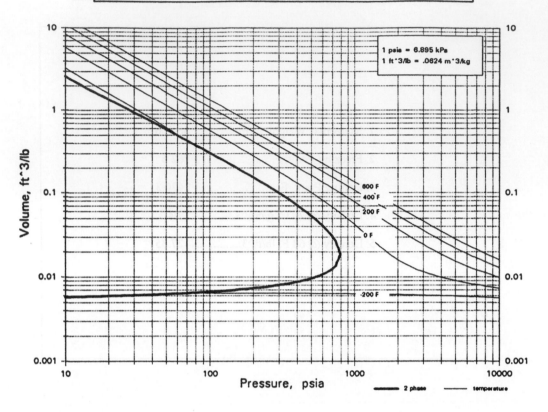

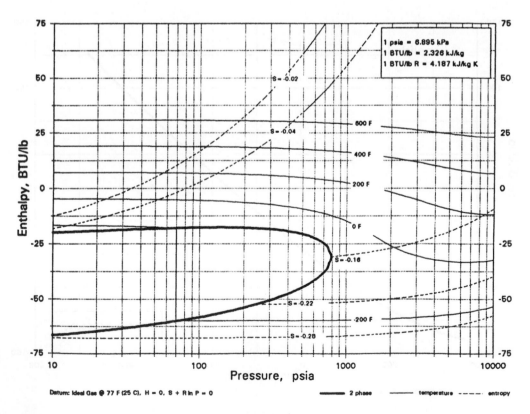

1. Molecular Weight, lb/mol.......... 138.906

2. Freezing Point, F....................... 1688

3. Boiling Point, F......................... 6097.7

4. Density @ 20 C, g/cm^3............ 6.15

5. Density @ 68 F, lb/ft^3.............. 383.93

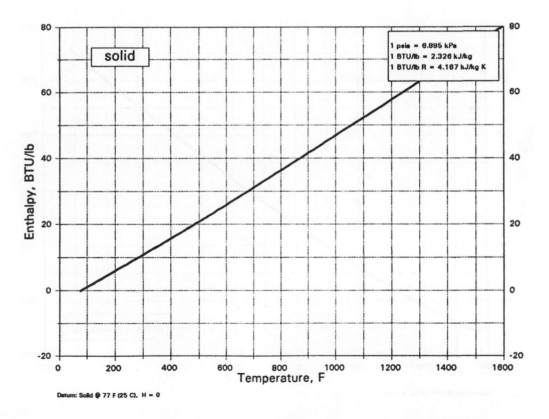

solid

1 psia = 6.895 kPa
1 BTU/lb = 2.326 kJ/kg
1 BTU/lb R = 4.187 kJ/kg K

Datum: Solid @ 77 F (25 C). H = 0

1. Molecular Weight, lb/mol.......... 6.941

2. Freezing Point, F....................... 356.9

3. Boiling Point, F......................... 2414.9

4. Density @ 20 C, g/cm^3............ 0.53

5. Density @ 68 F, lb/ft^3.............. 33.09

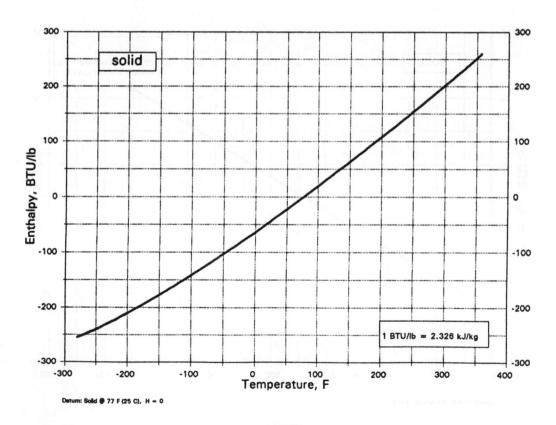

Datum: Solid @ 77 F (25 C), H = 0

1. Molecular Weight, lb/mol.......... 86.845

2. Freezing Point, F........................ 1016.6

3. Boiling Point, F.......................... 2390

4. Density @ 25 C, g/cm^3............ 3.46

5. Density @ 77 F, lb/ft^3.............. 216

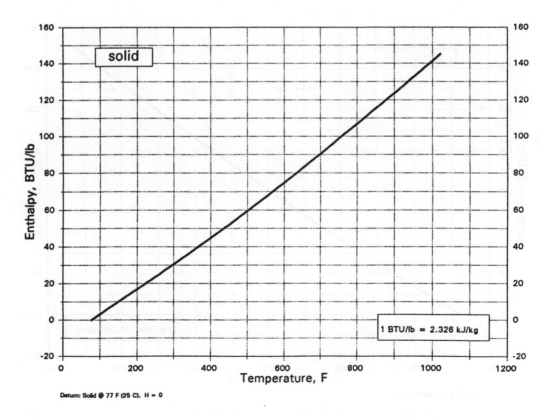

Datum: Solid @ 77 F (25 C), H = 0

155

1. Molecular Weight, lb/mol.......... 42.394

2. Freezing Point, F...................... 1137.2

3. Boiling Point, F........................ 2519.6

4. Density @ 25 C, g/cm^3............ 2.07

5. Density @ 77 F, lb/ft^3.............. 129.23

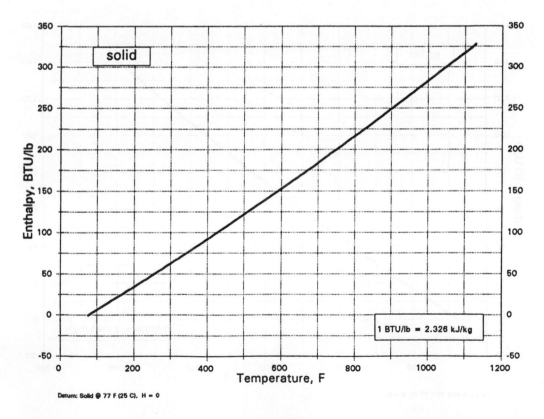

Datum: Solid @ 77 F (25 C), H = 0

1. Molecular Weight, lb/mol.......... 25.939

2. Freezing Point, F....................... 1598

3. Boiling Point, F.......................... 3057.8

4. Density @ 20 C, g/cm^3............ 2.64

5. Density @ 68 F, lb/ft^3.............. 164.81

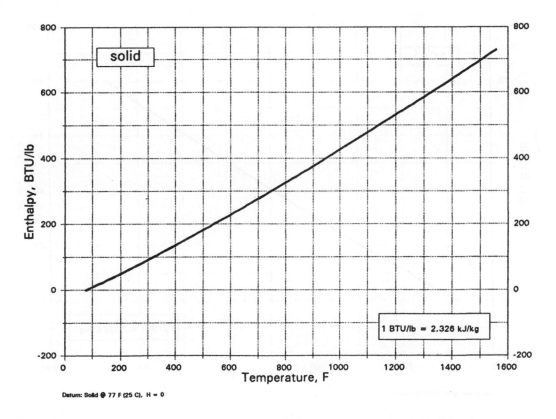

Datum: Solid @ 77 F (25 C), H = 0

1. Molecular Weight, lb/mol.......... 133.845

2. Freezing Point, F....................... 834.8

3. Boiling Point, F......................... 2139.8

4. Density @ 20 C, g/cm^3............ 4.08

5. Density @ 68 F, lb/ft^3.............. 254.71

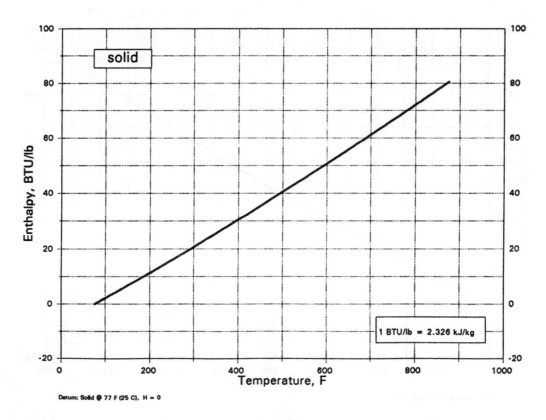

solid

1 BTU/lb = 2.326 kJ/kg

Enthalpy, BTU/lb

Temperature, F

Datum: Solid @ 77 F (25 C), H = 0

1. Molecular Weight, lb/mol.......... 174.967

2. Freezing Point, F........................ 3025.4

3. Boiling Point, F......................... 4103.3

4. Density @ 20 C, g/cm^3............ 9.84

5. Density @ 68 F, lb/ft^3.............. 614.29

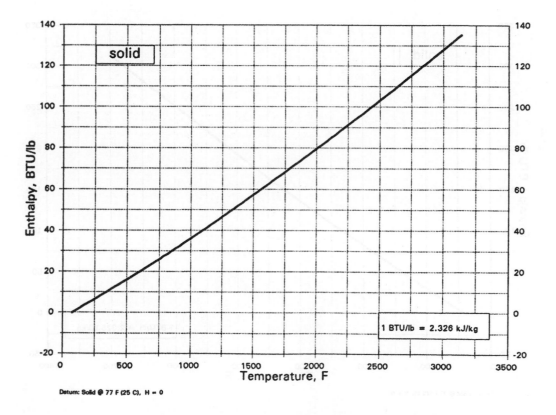

Datum: Solid @ 77 F (25 C), H = 0

159

1. Molecular Weight, lb/mol.......... 24.305

2. Freezing Point, F...................... 1202

3. Boiling Point, F........................ 2017.1

4. Density @ 20 C, g/cm^3............ 1.74

5. Density @ 68 F, lb/ft^3.............. 108.62

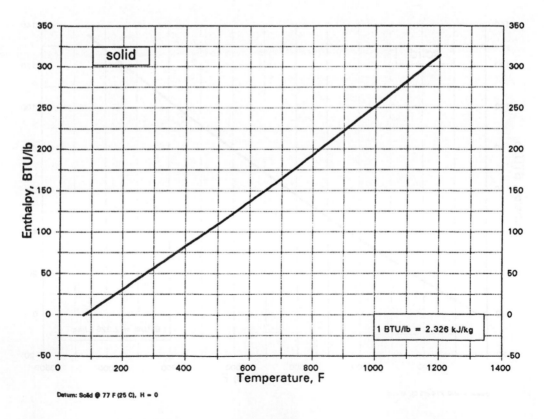

Datum: Solid @ 77 F (25 C), H = 0

1. Molecular Weight, lb/mol.......... 95.21

2. Freezing Point, F........................ 1313.6

3. Boiling Point, F......................... 2584.4

4. Density @ 20 C, g/cm^3............ 2.32

5. Density @ 68 F, lb/ft^3.............. 144.83

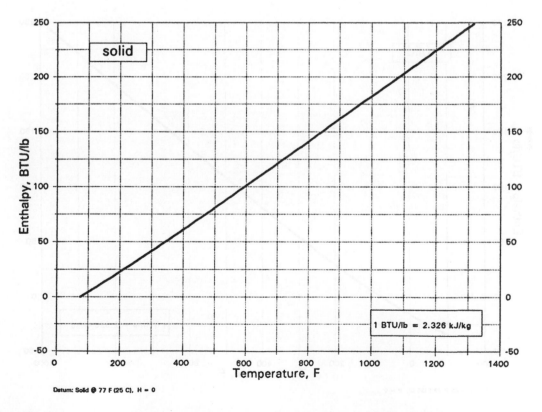

Datum: Solid @ 77 F (25 C), H = 0

161

1. Molecular Weight, lb/mol.......... 40.304

2. Freezing Point, F...................... 5129.3

3. Boiling Point, F........................ 6512.1

4. Density @ 25 C, g/cm^3............ 3.58

5. Density @ 77 F, lb/ft^3.............. 223.49

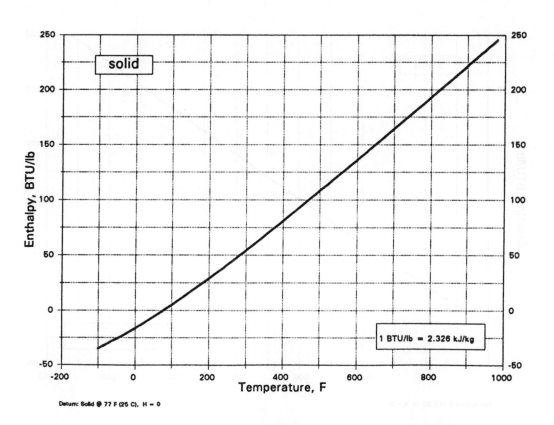

Datum: Solid @ 77 F (25 C), H = 0

1. Molecular Weight, lb/mol.......... 54.938

2. Freezing Point, F....................... 2274.8

3. Boiling Point, F......................... 3845.9

4. Density @ 20 C, g/cm^3............ 7.2

5. Density @ 68 F, lb/ft^3.............. 449.48

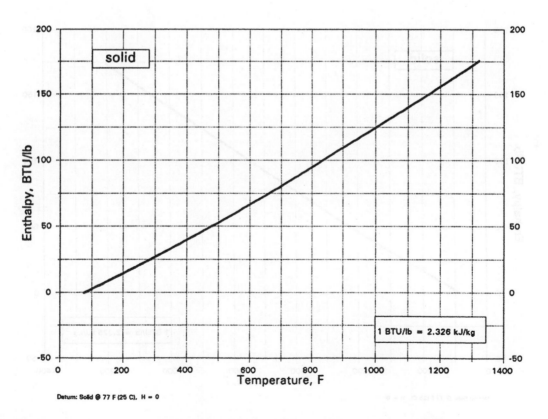

solid

Enthalpy, BTU/lb vs Temperature, F

1 BTU/lb = 2.326 kJ/kg

Datum: Solid @ 77 F (25 C), H = 0

1. Molecular Weight, lb/mol.......... 125.843

2. Freezing Point, F....................... 1202

3. Boiling Point, F......................... 2174

4. Density @ 25 C, g/cm^3............ 2.98

5. Density @ 77 F, lb/ft^3............. 186.03

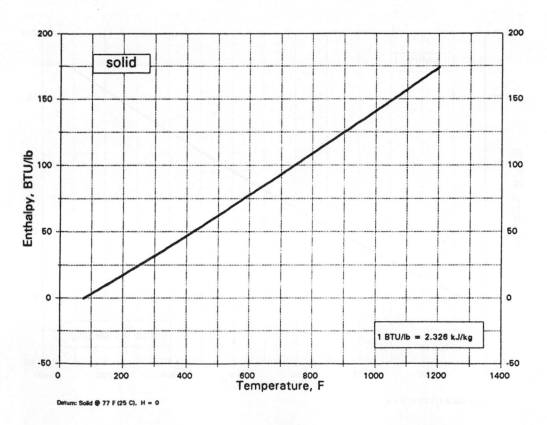

Datum: Solid @ 77 F (25 C), H = 0

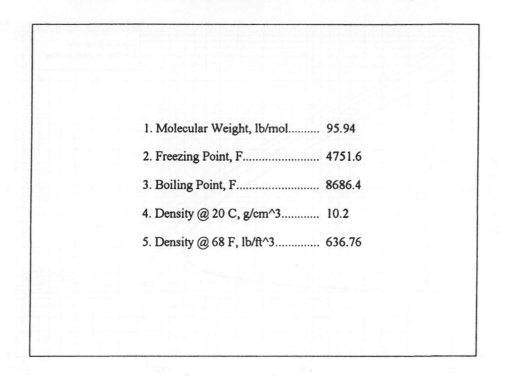

1. Molecular Weight, lb/mol.......... 95.94

2. Freezing Point, F........................ 4751.6

3. Boiling Point, F......................... 8686.4

4. Density @ 20 C, g/cm^3............ 10.2

5. Density @ 68 F, lb/ft^3.............. 636.76

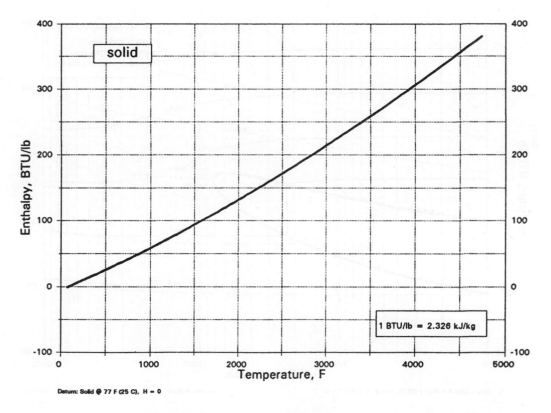

Datum: Solid @ 77 F (25 C), H = 0

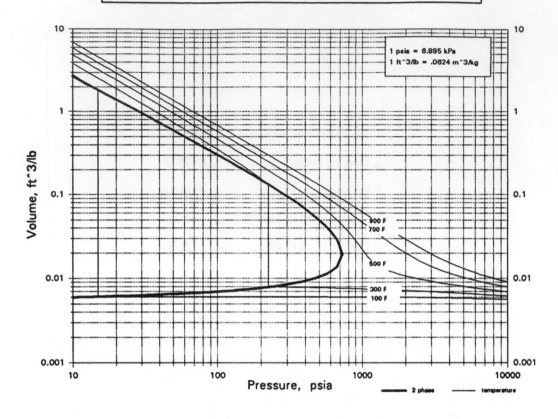

1 psia = 6.895 kPa
1 ft^3/lb = .0624 m^3/kg

900 F
700 F
500 F
300 F
100 F

Volume, ft^3/lb

Pressure, psia

2 phase   temperature

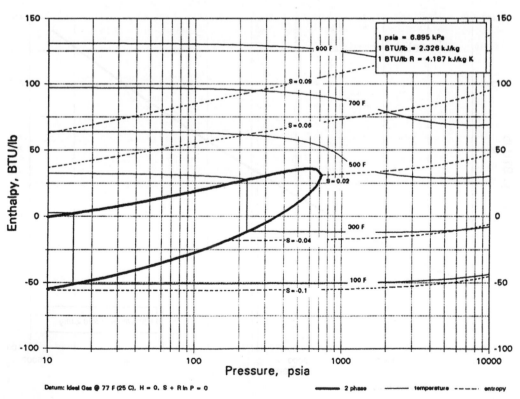

1 psia = 6.895 kPa
1 BTU/lb = 2.326 kJ/kg
1 BTU/lb R = 4.187 kJ/kg K

900 F
S = 0.09
700 F
S = 0.06
500 F
S = 0.02
300 F
S = -.04
100 F
S = -.1

Enthalpy, BTU/lb

Pressure, psia

Datum: Ideal Gas @ 77 F (25 C). H = 0. S + R ln P = 0

2 phase   temperature   entropy

1. Molecular Weight, lb/mol.......... 143.938

2. Freezing Point, F....................... 1463

3. Boiling Point, F......................... 2103.8

4. Density @ 21 C, g/cm^3............. 4.69

5. Density @ 70 F, lb/ft^3.............. 292.79

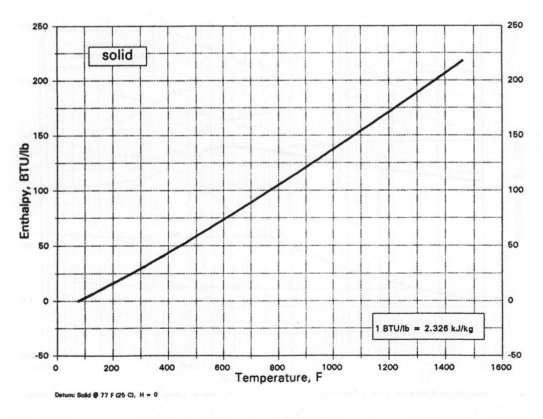

Datum: Solid @ 77 F (25 C). H = 0

167

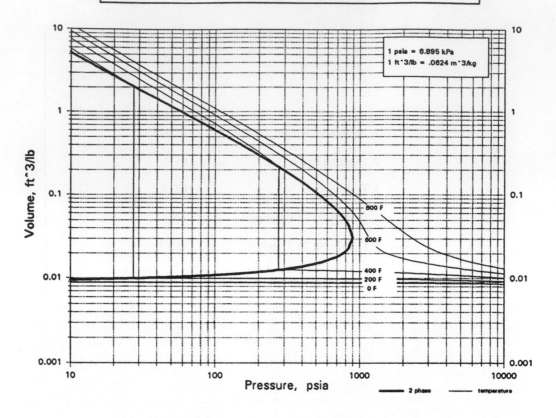

1 psia = 6.895 kPa
1 ft^3/lb = .0624 m^3/kg

800 F
600 F
400 F
200 F
0 F

Volume, ft^3/lb

Pressure, psia

2 phase   temperature

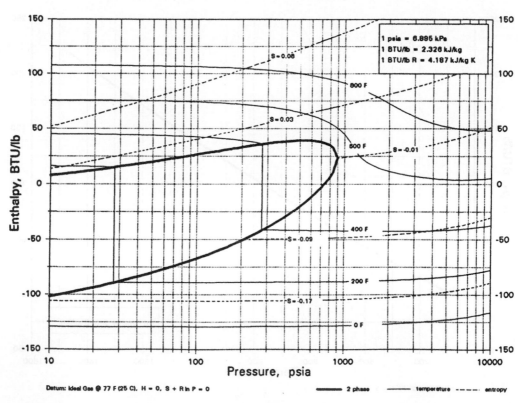

1 psia = 6.895 kPa
1 BTU/lb = 2.326 kJ/kg
1 BTU/lb R = 4.187 kJ/kg K

S = 0.08
S = 0.03
S = -0.01
S = -0.09
S = -0.17

800 F
600 F
400 F
200 F
0 F

Enthalpy, BTU/lb

Pressure, psia

Datum: Ideal Gas @ 77 F (25 C). H = 0. S + R ln P = 0

2 phase   temperature   entropy

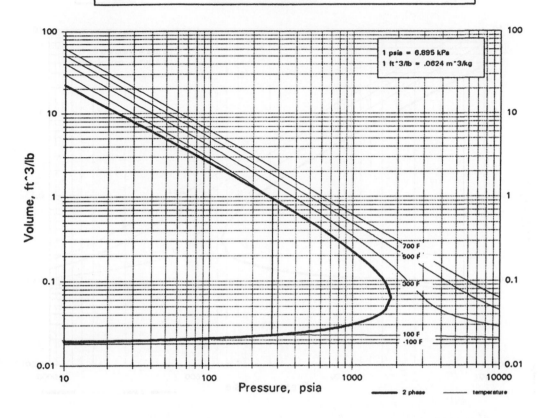

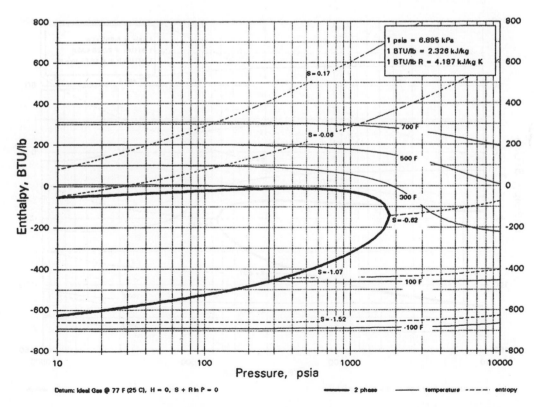

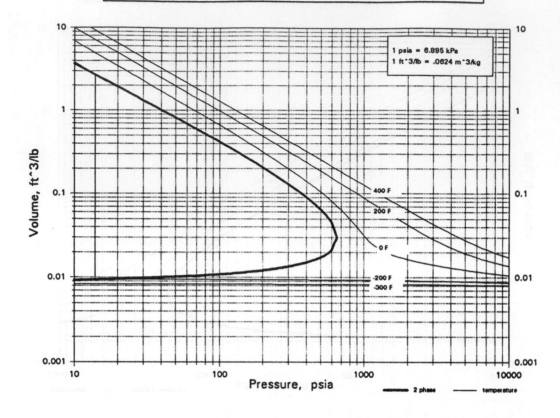

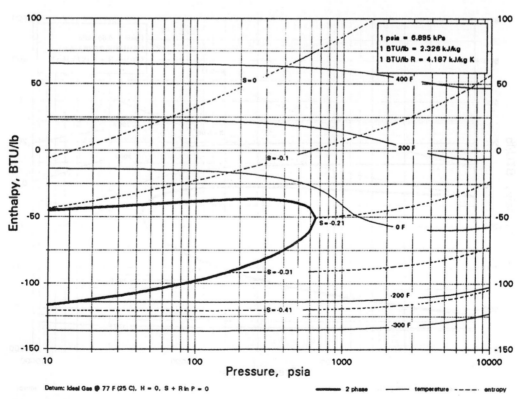

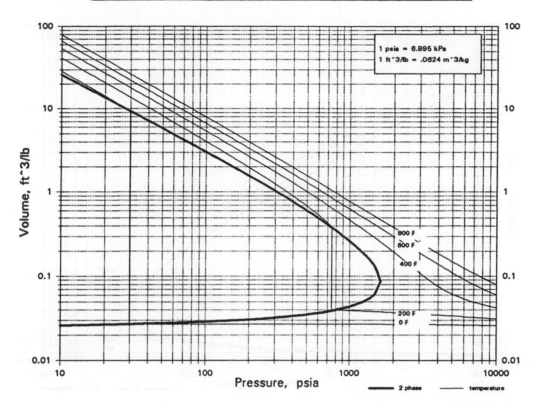

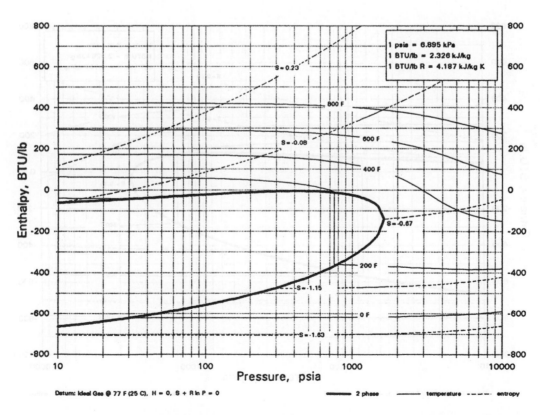

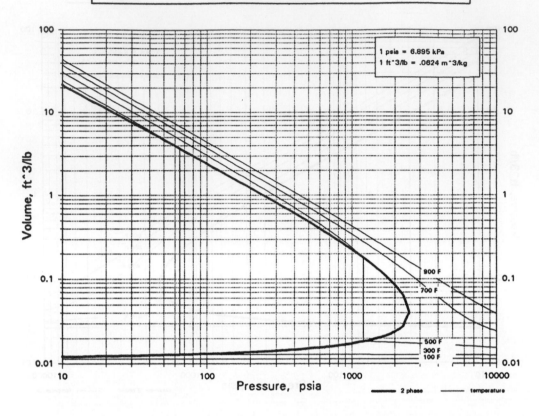

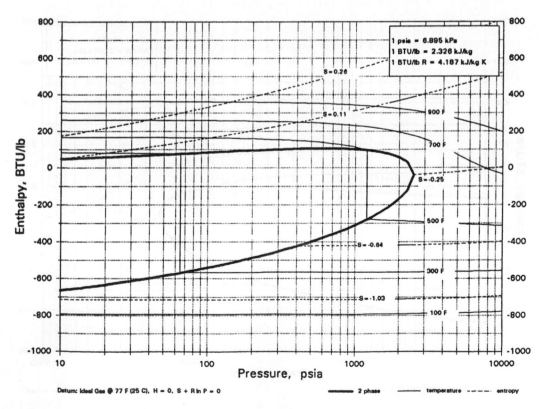

NH4Br	AMMONIUM BROMIDE

1. Molecular Weight, lb/mol.......... 97.943

2. Freezing Point, F........................ ---

3. Boiling Point, F.......................... 744.8

4. Density @ 20 C, g/cm^3............ 2.43

5. Density @ 68 F, lb/ft^3.............. 151.7

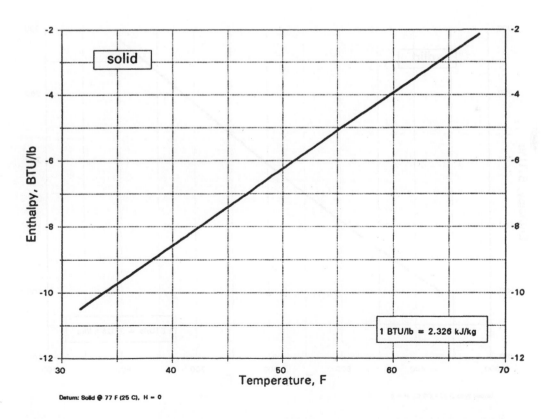

solid

1 BTU/lb = 2.326 kJ/kg

Datum: Solid @ 77 F (25 C), H = 0

1. Molecular Weight, lb/mol.......... 53.491

2. Freezing Point, F....................... 968.1

3. Boiling Point, F......................... 641.9

4. Density @ 20 C, g/cm^3............. 1.53

5. Density @ 68 F, lb/ft^3.............. 95.51

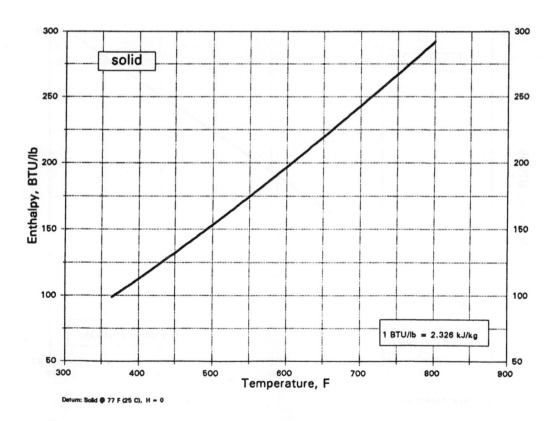

Datum: Solid @ 77 F (25 C), H = 0

1. Molecular Weight, lb/mol.......... 144.943

2. Freezing Point, F....................... ---

3. Boiling Point, F....................... 760.8

4. Density @ 25 C, g/cm^3............ 2.51

5. Density @ 77 F, lb/ft^3.............. 156.69

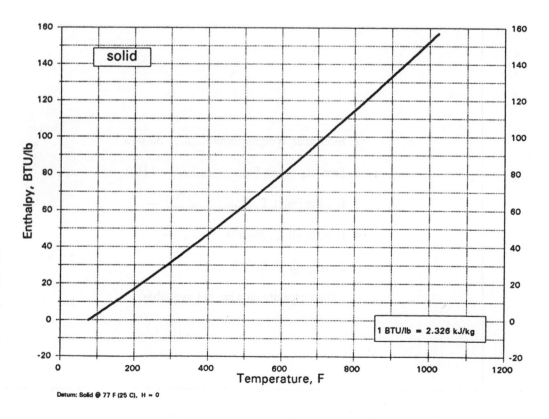

Datum: Solid @ 77 F (25 C), H = 0

1. Molecular Weight, lb/mol.......... 35.046

2. Freezing Point, F........................ -110.2

3. Boiling Point, F......................... ---

4. Density @ C, g/cm^3............. ---

5. Density @ F, lb/ft^3.............. ---

---

1. Molecular Weight, lb/mol.......... 35.046

2. Freezing Point, F........................ -110.2

3. Boiling Point, F......................... ---

4. Density @ C, g/cm^3............. ---

5. Density @ F, lb/ft^3.............. ---

Heat capacity data are not available.

1. Molecular Weight, lb/mol.......... 51.112

2. Freezing Point, F........................ 244.4

3. Boiling Point, F........................... 91.9

4. Density @ 20 C, g/cm^3............ 1.17

5. Density @ 68 F, lb/ft^3.............. 73.04

1. Molecular Weight, lb/mol.......... 51.112

2. Freezing Point, F........................ 244.4

3. Boiling Point, F........................... 91.9

4. Density @ 20 C, g/cm^3............ 1.17

5. Density @ 68 F, lb/ft^3.............. 73.04

Heat capacity data are not available.

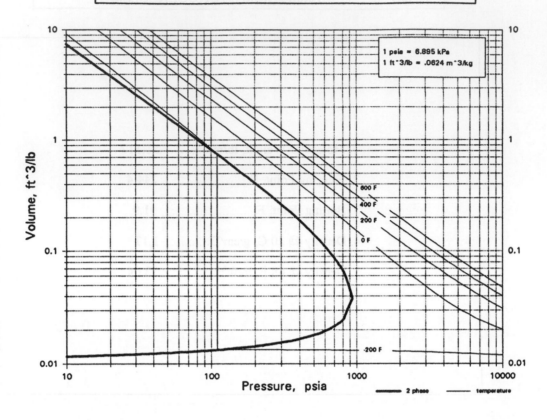

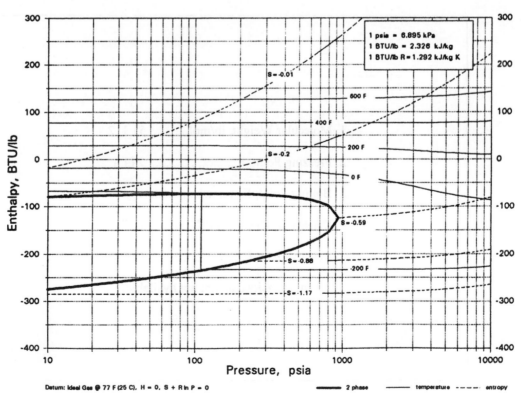

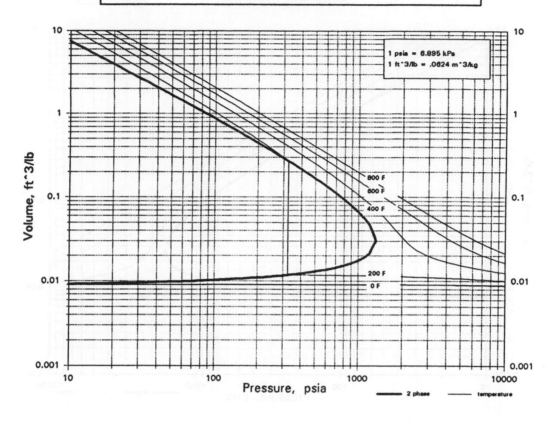

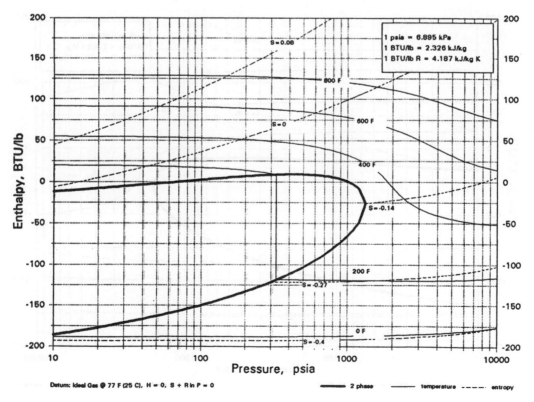

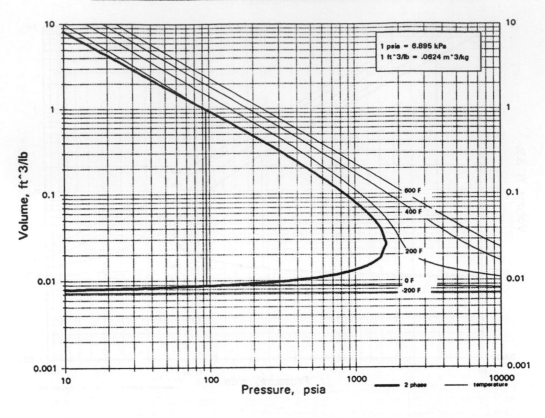

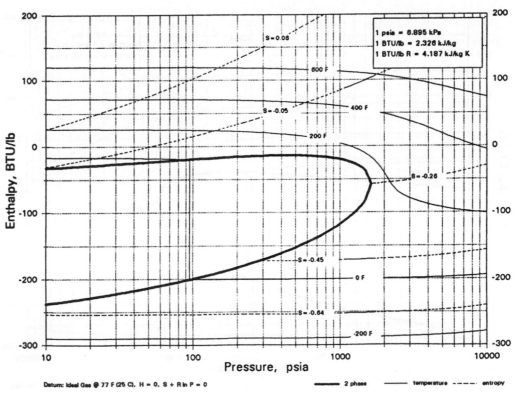

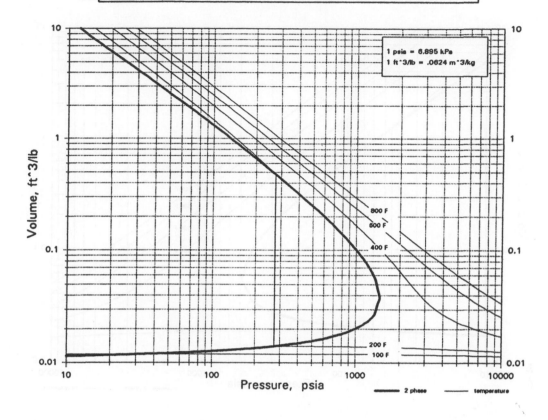

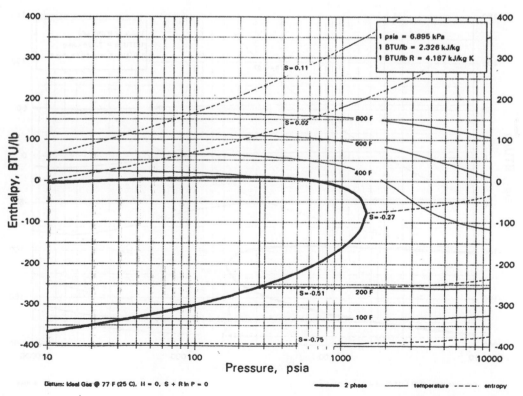

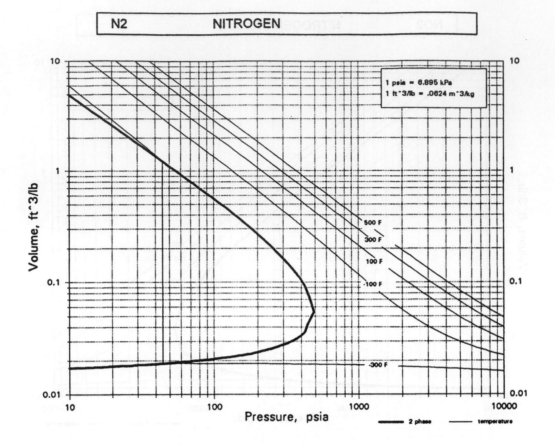

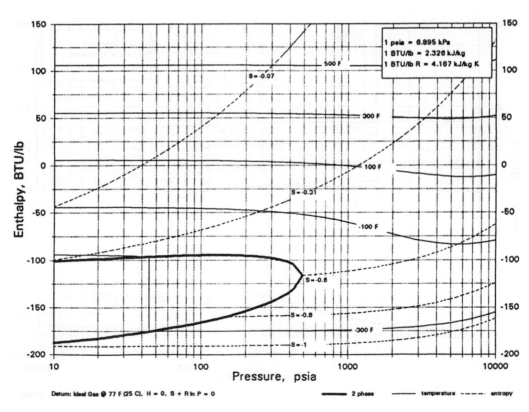

Datum: Ideal Gas @ 77 F (25 C), H = 0, S + R ln P = 0

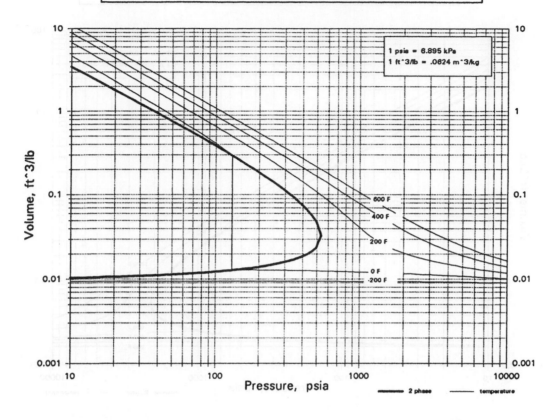

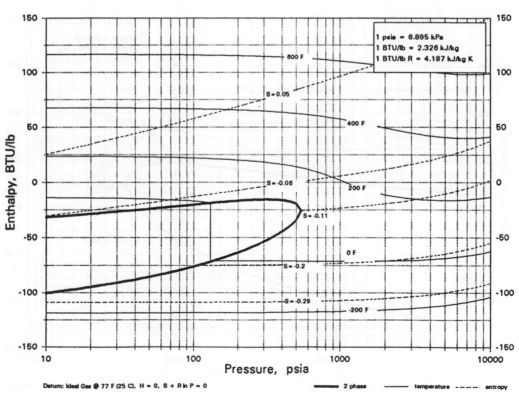

Datum: Ideal Gas @ 77 F (25 C), H = 0, S + R ln P = 0

183

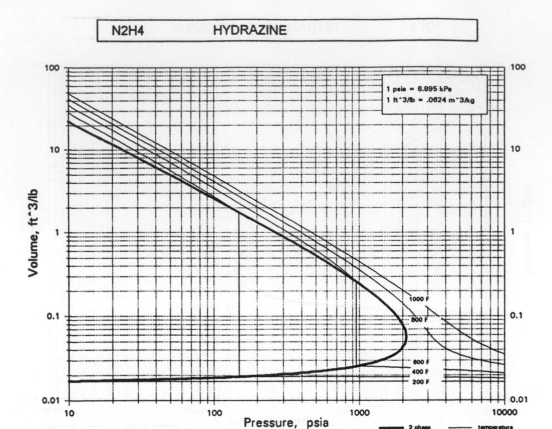

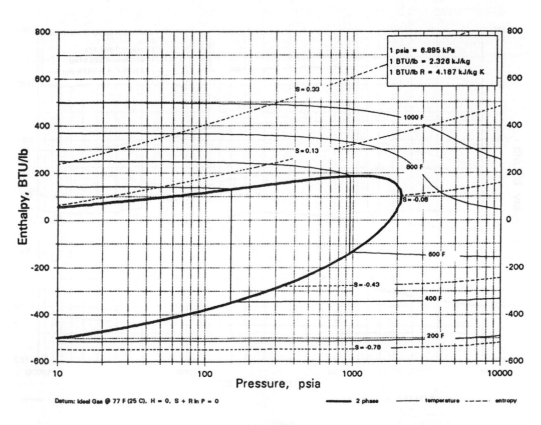

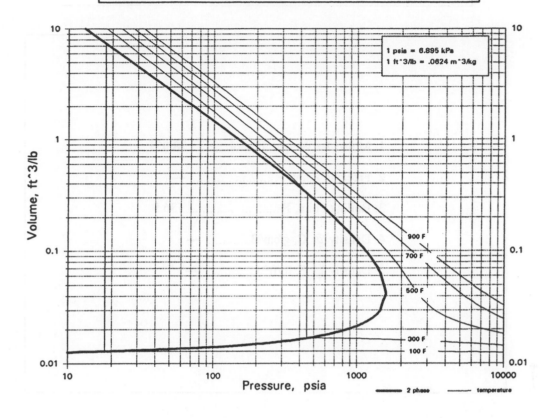

1 psia = 6.895 kPa
1 ft^3/lb = .0624 m^3/kg

900 F
700 F
500 F
300 F
100 F

Volume, ft^3/lb

Pressure, psia

2 phase   temperature

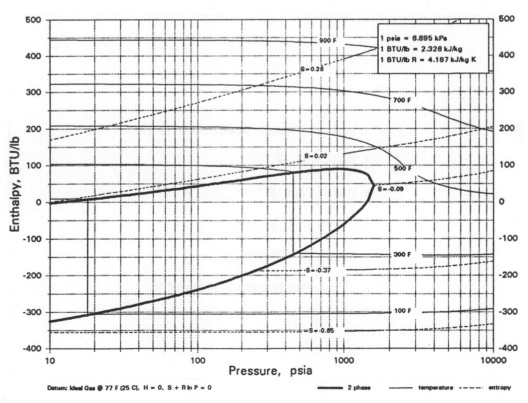

1 psia = 6.895 kPa
1 BTU/lb = 2.326 kJ/kg
1 BTU/lb R = 4.187 kJ/kg K

900 F
S = 0.26
700 F
S = 0.02
500 F
S = -0.09
300 F
S = -0.37
100 F
S = -0.65

Enthalpy, BTU/lb

Pressure, psia

Datum: Ideal Gas @ 77 F (25 C), H = 0, S + R ln P = 0

2 phase   temperature   ----- entropy

185

1. Molecular Weight, lb/mol.......... 78.071

2. Freezing Point, F...................... ---

3. Boiling Point, F......................... 136.9

4. Density @ C, g/cm^3............. ---

5. Density @ F, lb/ft^3.............. ---

1. Molecular Weight, lb/mol.......... 78.071

2. Freezing Point, F...................... ---

3. Boiling Point, F......................... 136.9

4. Density @ C, g/cm^3............. ---

5. Density @ F, lb/ft^3.............. ---

Heat capacity data are not available.

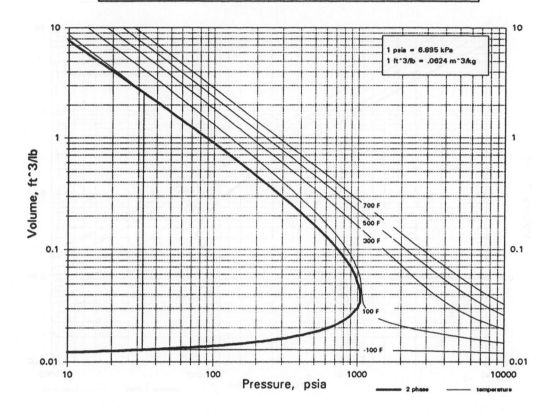

1 psia = 6.895 kPa
1 ft^3/lb = .0624 m^3/kg

700 F
500 F
300 F
100 F
-100 F

— 2 phase   — temperature

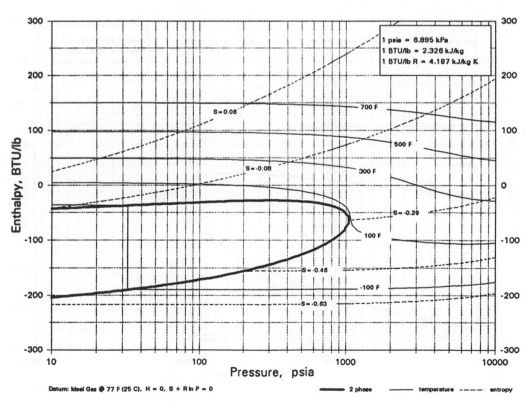

1 psia = 6.895 kPa
1 BTU/lb = 2.326 kJ/kg
1 BTU/lb R = 4.187 kJ/kg K

S = 0.06
S = -0.08
700 F
500 F
300 F
S = -0.29
100 F
S = -0.46
-100 F
S = -0.63

Datum: Ideal Gas @ 77 F (25 C), H = 0, S + R ln P = 0

— 2 phase   — temperature   --- entropy

187

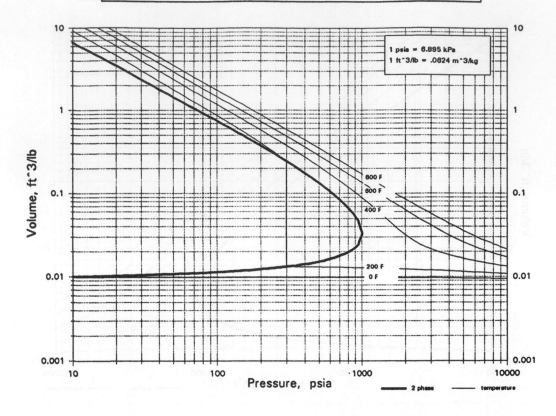

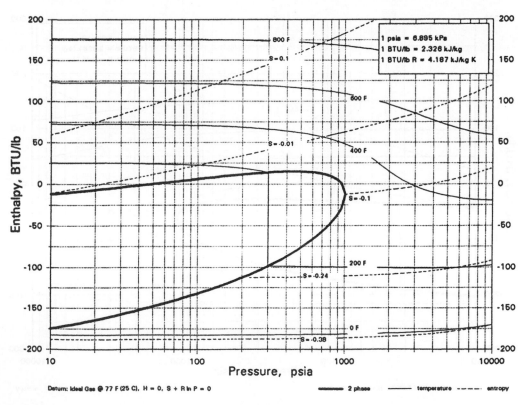

188

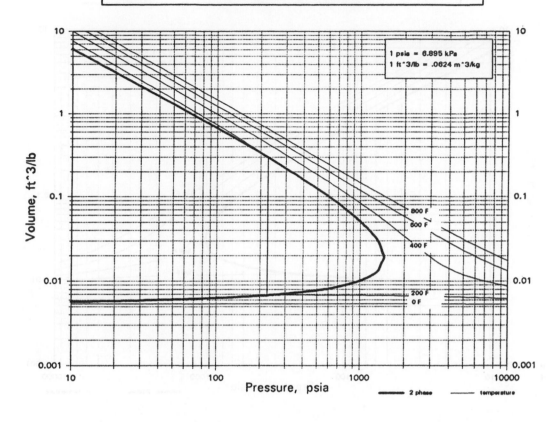

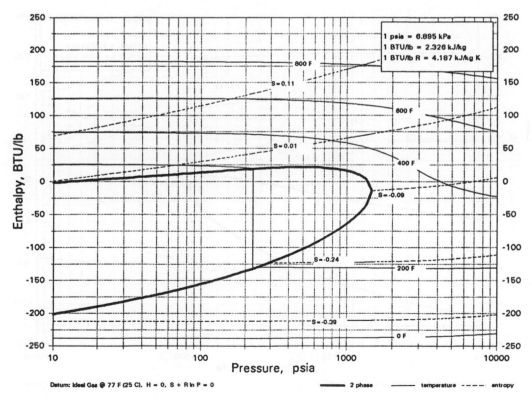

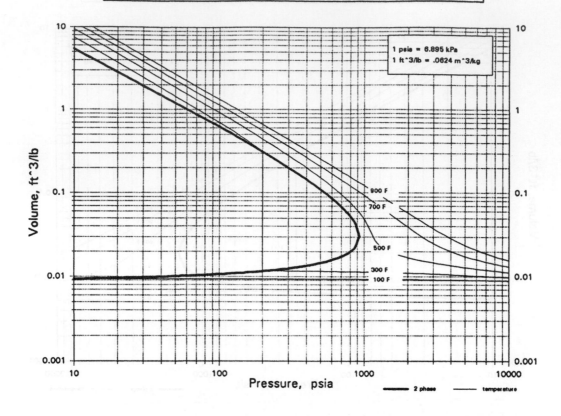

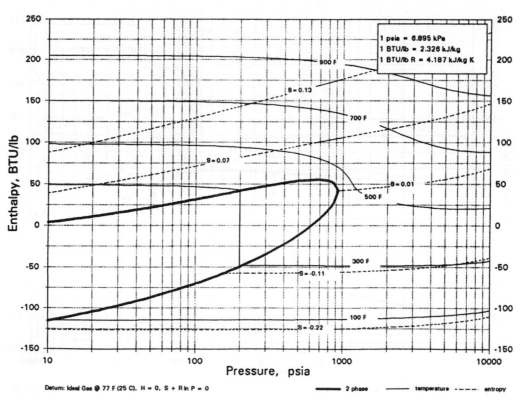

1. Molecular Weight, lb/mol.......... 22.99

2. Freezing Point, F........................ 208.1

3. Boiling Point, F........................ 1621.1

4. Density @ 20 C, g/cm^3............ 0.97

5. Density @ 68 F, lb/ft^3.............. 60.55

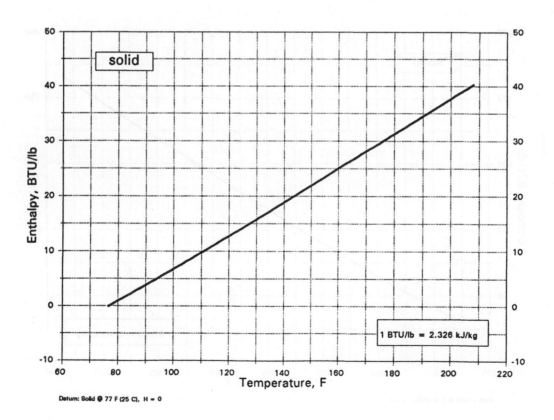

Datum: Solid @ 77 F (25 C), H = 0

1. Molecular Weight, lb/mol.......... 102.894

2. Freezing Point, F...................... 1376.3

3. Boiling Point, F......................... 2535.2

4. Density @ 25 C, g/cm^3............ 3.2

5. Density @ 77 F, lb/ft^3.............. 199.77

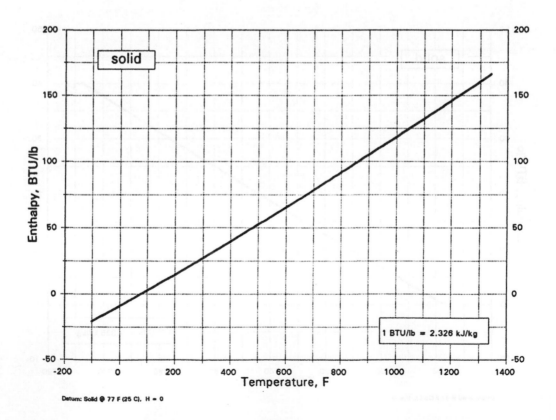

Datum: Solid @ 77 F (25 C). H = 0

1. Molecular Weight, lb/mol.......... 49.008

2. Freezing Point, F........................ 1046.6

3. Boiling Point, F......................... 2724.8

4. Density @ 20 C, g/cm^3............ 1.6

5. Density @ 68 F, lb/ft^3.............. 99.89

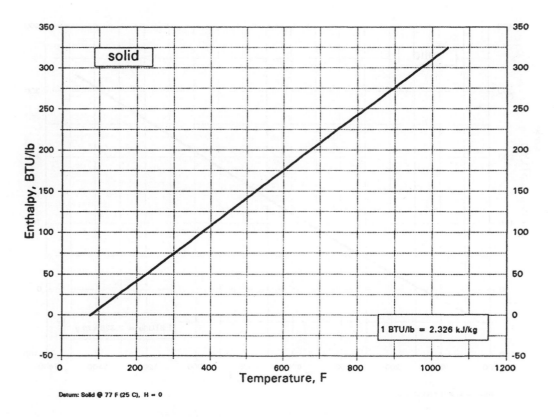

solid

1 BTU/lb = 2.326 kJ/kg

Datum: Solid @ 77 F (25 C), H = 0

1. Molecular Weight, lb/mol.......... 58.442

2. Freezing Point, F...................... 1473.4

3. Boiling Point, F........................ 2669

4. Density @ 25 C, g/cm^3............ 2.17

5. Density @ 77 F, lb/ft^3.............. 135.47

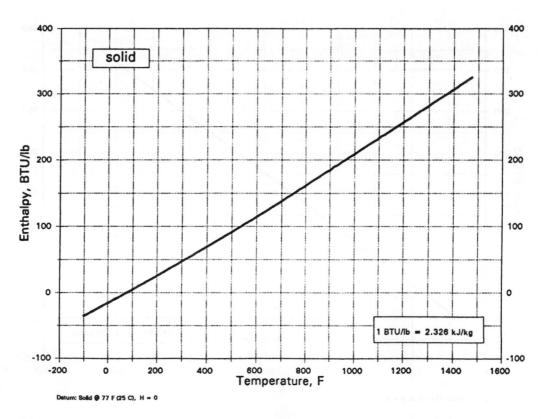

solid

1 BTU/lb = 2.326 kJ/kg

Datum: Solid @ 77 F (25 C), H = 0

194

1. Molecular Weight, lb/mol.......... 41.988

2. Freezing Point, F........................ 1824.5

3. Boiling Point, F........................ 3109.2

4. Density @ 41 C, g/cm^3............ 2.56

5. Density @ 106 F, lb/ft^3............ 159.82

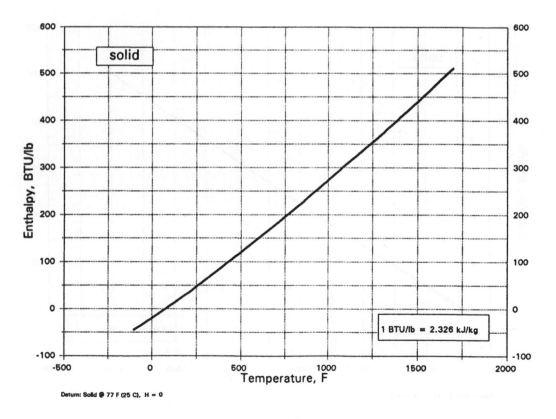

solid

1 BTU/lb = 2.326 kJ/kg

Datum: Solid @ 77 F (25 C), H = 0

1. Molecular Weight, lb/mol.......... 149.894

2. Freezing Point, F...................... 1203.8

3. Boiling Point, F......................... 2379.2

4. Density @ 25 C, g/cm^3............ 3.67

5. Density @ 77 F, lb/ft^3.............. 229.11

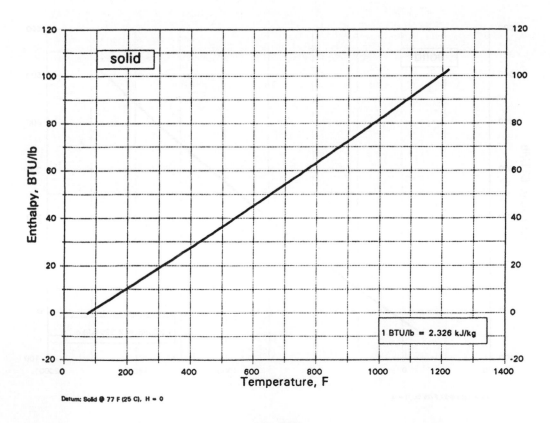

solid

1 BTU/lb = 2.326 kJ/kg

Enthalpy, BTU/lb

Temperature, F

Datum: Solid @ 77 F (25 C), H = 0

1. Molecular Weight, lb/mol.......... 39.997

2. Freezing Point, F...................... 613.1

3. Boiling Point, F........................ 2534

4. Density @ 20 C, g/cm^3............ 2.13

5. Density @ 68 F, lb/ft^3.............. 132.97

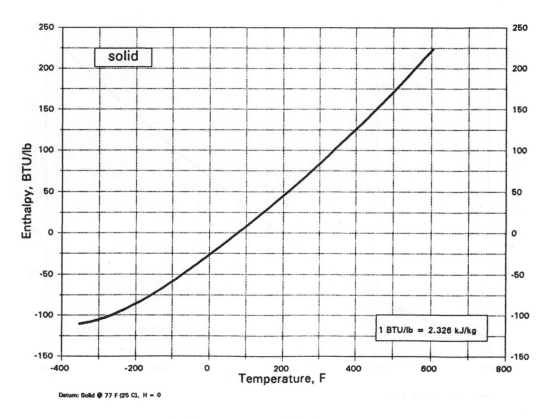

Datum: Solid @ 77 F (25 C), H = 0

1. Molecular Weight, lb/mol.......... 142.043

2. Freezing Point, F....................... 1622.9

3. Boiling Point, F......................... ---

4. Density @ 20 C, g/cm^3............ 2.68

5. Density @ 68 F, lb/ft^3.............. 167.31

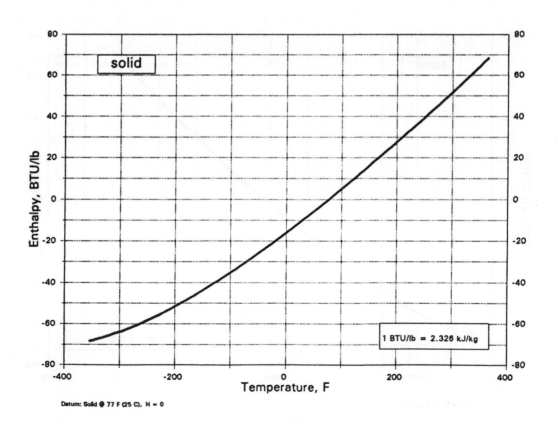

Datum: Solid @ 77 F (25 C), H = 0

1. Molecular Weight, lb/mol.......... 92.906

2. Freezing Point, F........................ 4490.6

3. Boiling Point, F.......................... 8747.3

4. Density @ 20 C, g/cm^3............ 8.57

5. Density @ 68 F, lb/ft^3.............. 535.01

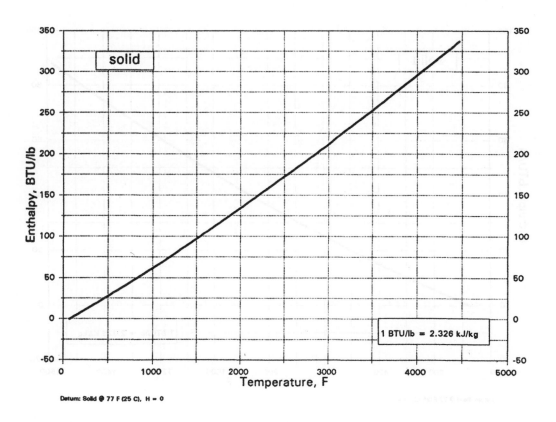

solid

1 BTU/lb = 2.326 kJ/kg

Datum: Solid @ 77 F (25 C), H = 0

199

1. Molecular Weight, lb/mol.......... 144.24

2. Freezing Point, F...................... 1860.8

3. Boiling Point, F......................... 5631.5

4. Density @ 20 C, g/cm^3............ 7.01

5. Density @ 68 F, lb/ft^3.............. 437.62

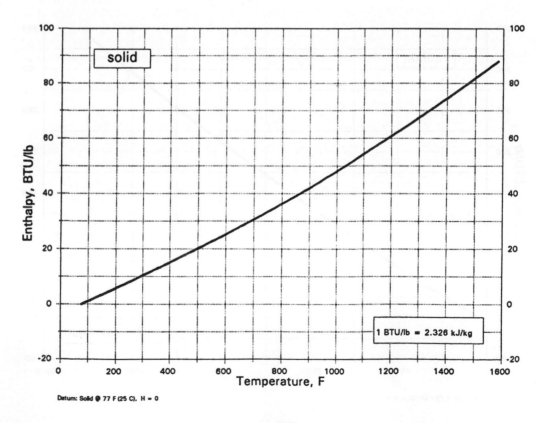

solid

1 BTU/lb = 2.326 kJ/kg

Datum: Solid @ 77 F (25 C), H = 0

Ne   NEON

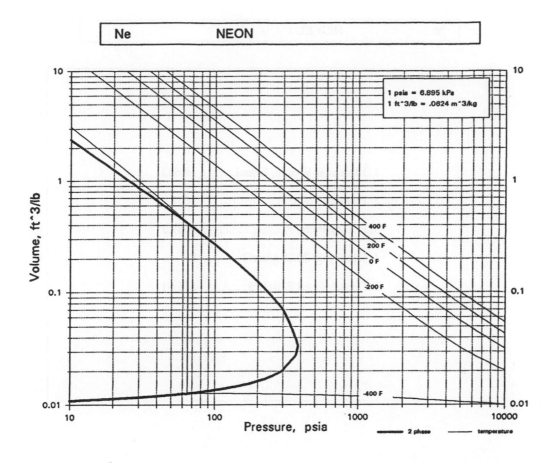

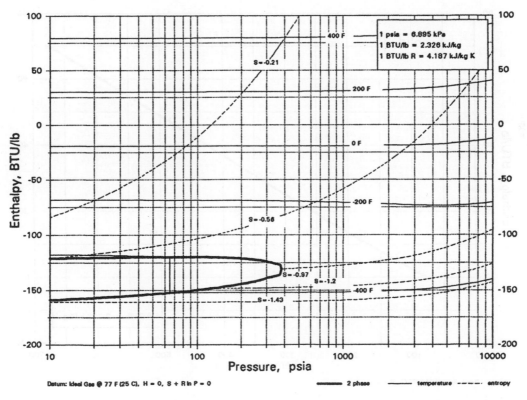

Datum: Ideal Gas @ 77 F (25 C), H = 0, S + R ln P = 0

201

1. Molecular Weight, lb/mol.......... 58.693

2. Freezing Point, F........................ 2651

3. Boiling Point, F......................... 3887.3

4. Density @ 20 C, g/cm^3............. 8.9

5. Density @ 68 F, lb/ft^3.............. 555.61

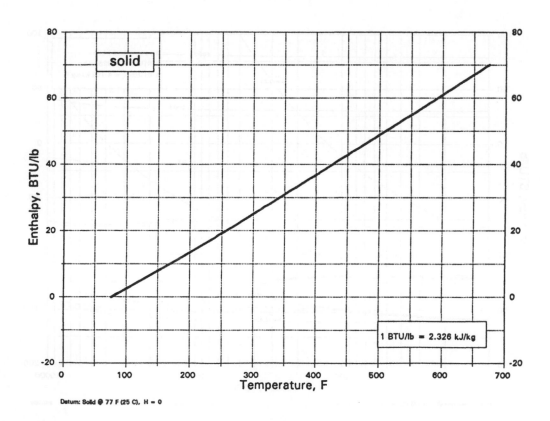

solid

1 BTU/lb = 2.326 kJ/kg

Datum: Solid @ 77 F (25 C), H = 0

202

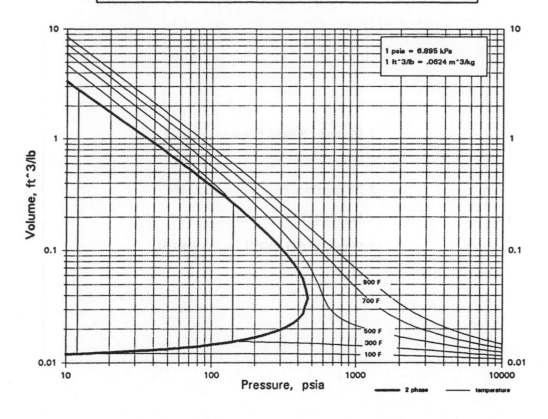

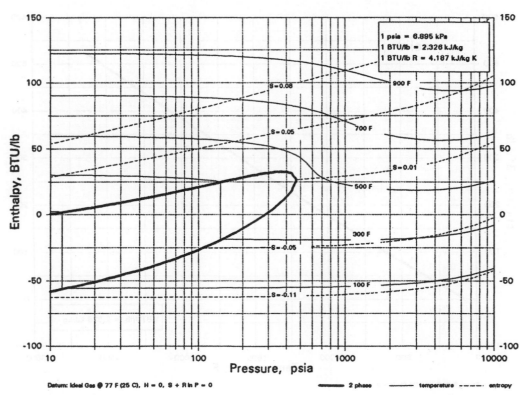

1. Molecular Weight, lb/mol.......... 96.69

2. Freezing Point, F...................... 2642

3. Boiling Point, F......................... 3164

4. Density @ 20 C, g/cm^3........... 4.63

5. Density @ 68 F, lb/ft^3.............. 289.04

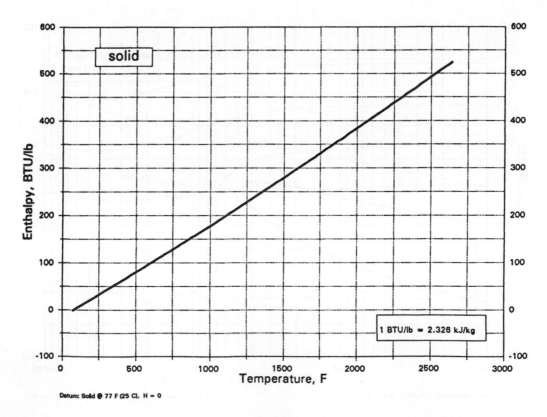

solid

1 BTU/lb = 2.326 kJ/kg

Datum: Solid @ 77 F (25 C), H = 0

1. Molecular Weight, lb/mol.......... 237

2. Freezing Point, F....................... 1184

3. Boiling Point, F.......................... 7055.6

4. Density @ 20 C, g/cm^3............ 20.45

5. Density @ 68 F, lb/ft^3.............. 1276.65

1. Molecular Weight, lb/mol.......... 237

2. Freezing Point, F....................... 1184

3. Boiling Point, F.......................... 7055.6

4. Density @ 20 C, g/cm^3............ 20.45

5. Density @ 68 F, lb/ft^3.............. 1276.65

Heat capacity data are not available.

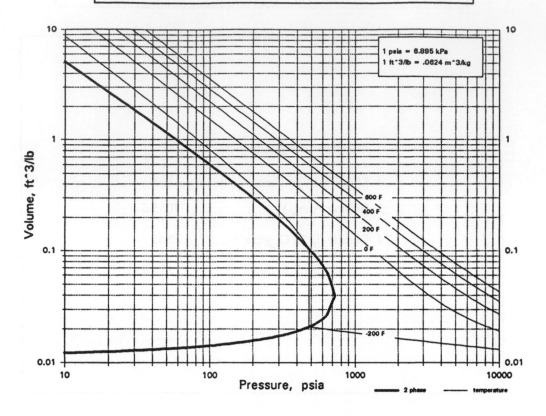

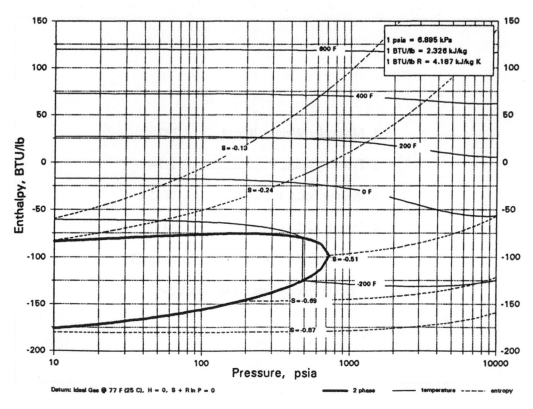

Datum: Ideal Gas @ 77 F (25 C), H = 0, S + R ln P = 0   2 phase   temperature   entropy

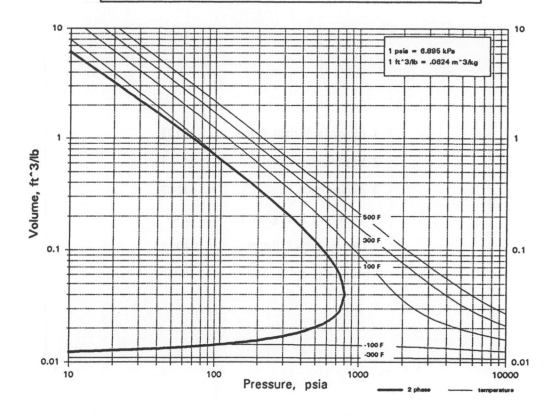

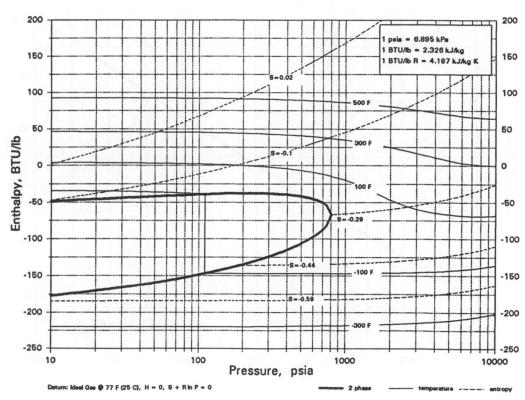

1. Molecular Weight, lb/mol.......... 190.23

2. Freezing Point, F...................... 5491.4

3. Boiling Point, F......................... 8324.3

4. Density @ 20 C, g/cm^3........... 22.48

5. Density @ 68 F, lb/ft^3............. 1403.38

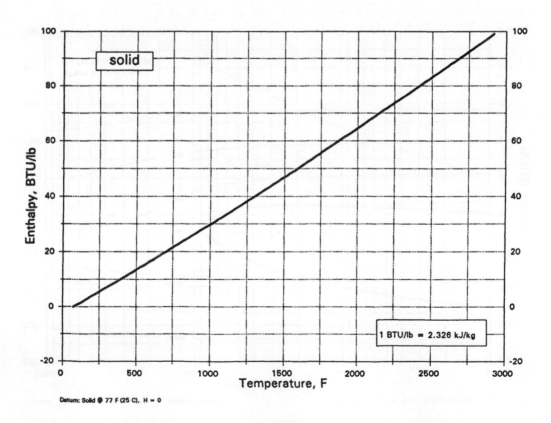

Datum: Solid @ 77 F (25 C), H = 0

1. Molecular Weight, lb/mol.......... 301.221

2. Freezing Point, F........................ 139.6

3. Boiling Point, F........................ 212.9

4. Density @ C, g/cm^3............ ---

5. Density @ F, lb/ft^3.............. ---

1. Molecular Weight, lb/mol.......... 301.221

2. Freezing Point, F........................ 139.6

3. Boiling Point, F........................ 212.9

4. Density @ C, g/cm^3............ ---

5. Density @ F, lb/ft^3.............. ---

Heat capacity data are not available.

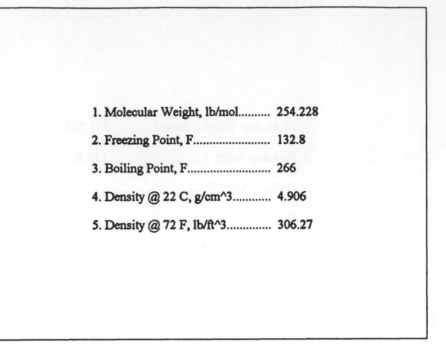

1. Molecular Weight, lb/mol.......... 254.228

2. Freezing Point, F........................ 132.8

3. Boiling Point, F........................... 266

4. Density @ 22 C, g/cm^3............. 4.906

5. Density @ 72 F, lb/ft^3............... 306.27

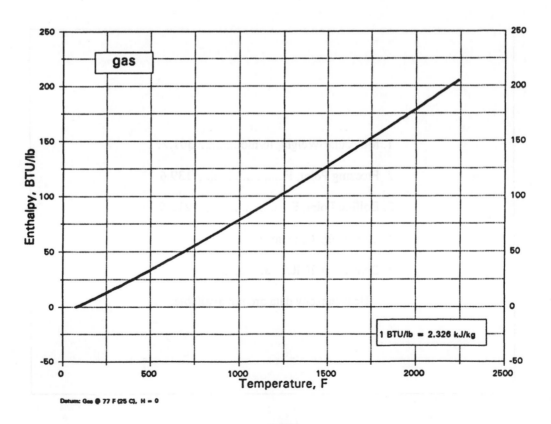

gas

1 BTU/lb = 2.326 kJ/kg

Datum: Gas @ 77 F (25 C), H = 0

1. Molecular Weight, lb/mol.......... 254.228

2. Freezing Point, F...................... 107.6

3. Boiling Point, F......................... 266

4. Density @ 22 C, g/cm^3............. 4.906

5. Density @ 72 F, lb/ft^3.............. 306.27

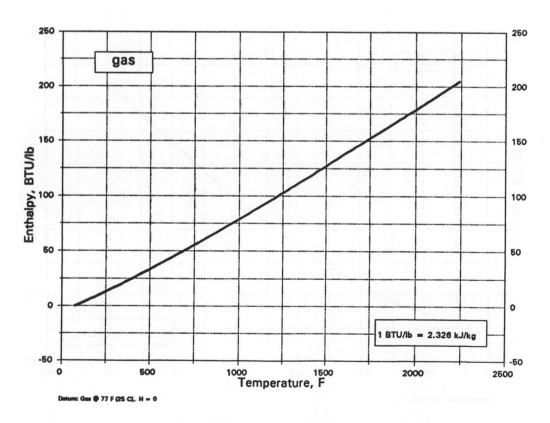

gas

Enthalpy, BTU/lb

Temperature, F

1 BTU/lb = 2.326 kJ/kg

Datum: Gas @ 77 F (25 C), H = 0

211

1. Molecular Weight, lb/mol.......... 30.974

2. Freezing Point, F....................... 111.38

3. Boiling Point, F......................... 536.5

4. Density @ 20 C, g/cm^3............. 1.82

5. Density @ 68 F, lb/ft^3.............. 113.62

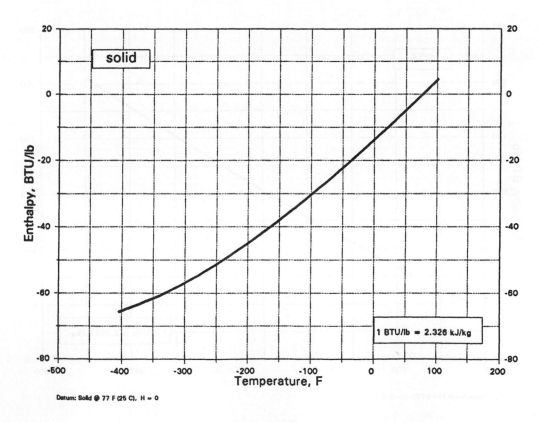

solid

1 BTU/lb = 2.326 kJ/kg

Datum: Solid @ 77 F (25 C), H = 0

212

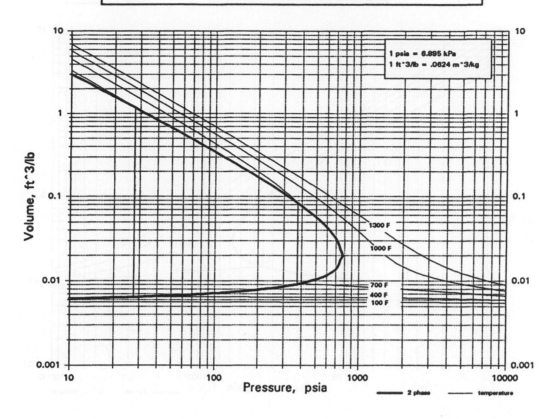

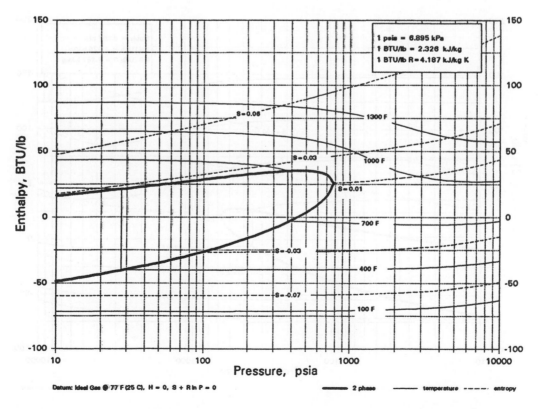

Datum: Ideal Gas @ 77 F (25 C), H = 0, S + R ln P = 0   2 phase   temperature   entropy

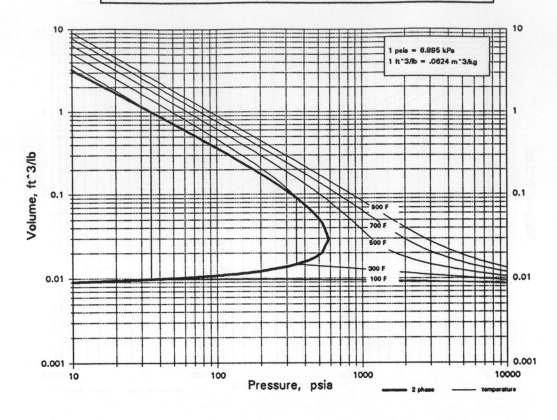

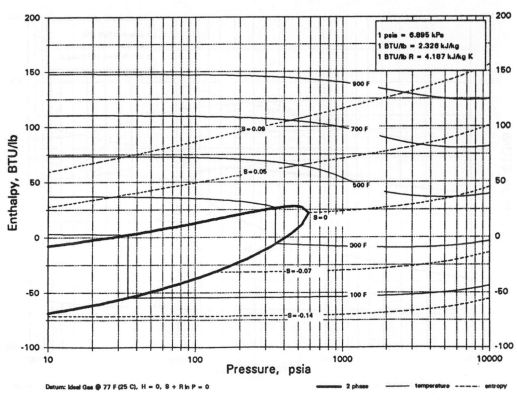

Datum: Ideal Gas @ 77 F (25 C), H = 0, S + R ln P = 0

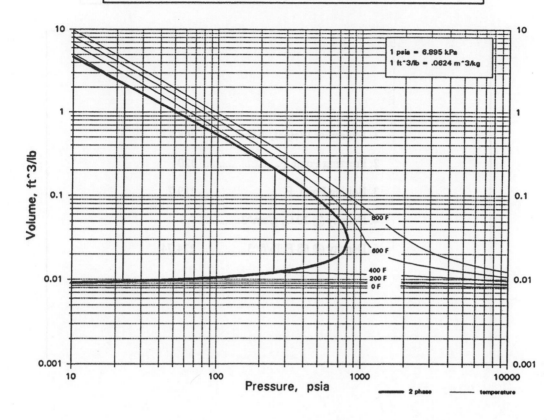

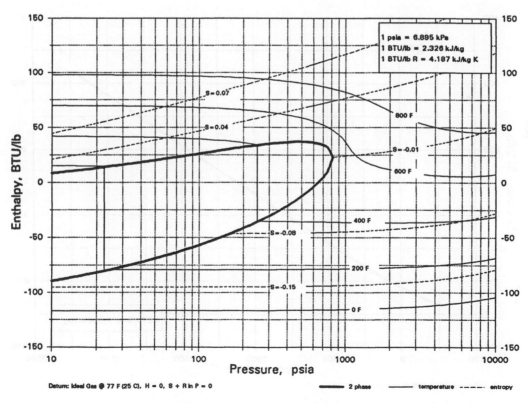

Datum: Ideal Gas @ 77 F (25 C), H = 0, S + R ln P = 0

1. Molecular Weight, lb/mol.......... 208.237

2. Freezing Point, F........................ 320

3. Boiling Point, F.......................... 319.7

4. Density @ 20 C, g/cm^3............. 3.6

5. Density @ 68 F, lb/ft^3.............. 224.74

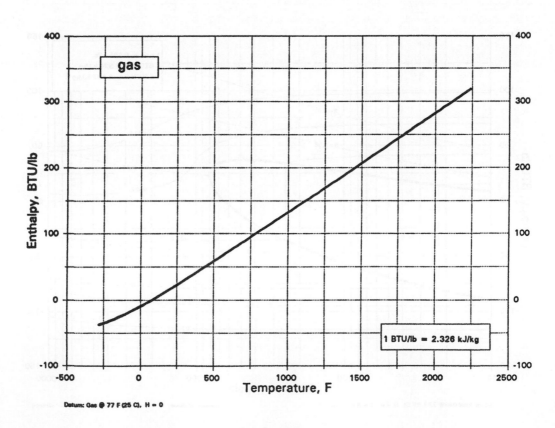

Datum: Gas @ 77 F (25 C), H = 0

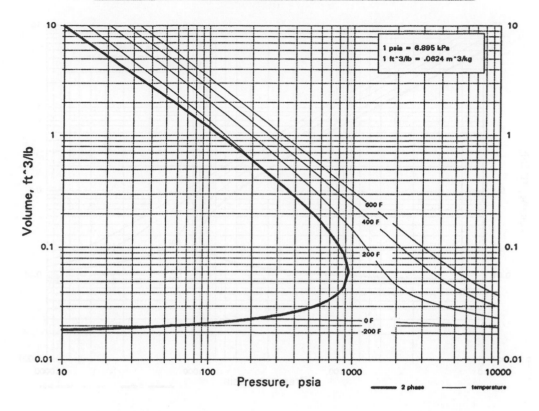

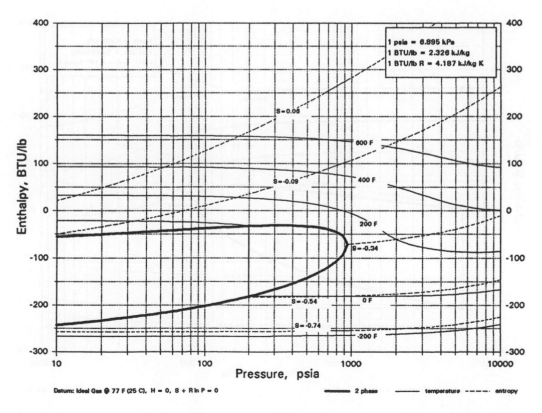

217

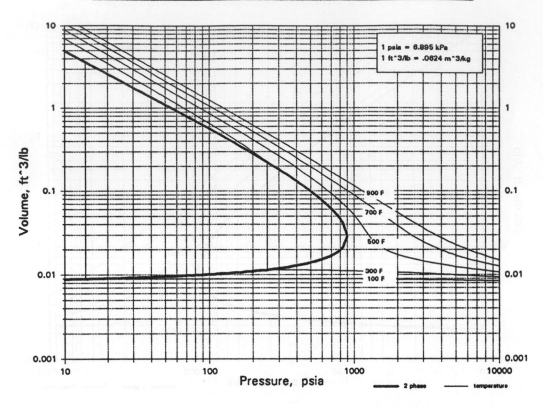

1 psia = 6.895 kPa
1 ft^3/lb = .0624 m^3/kg

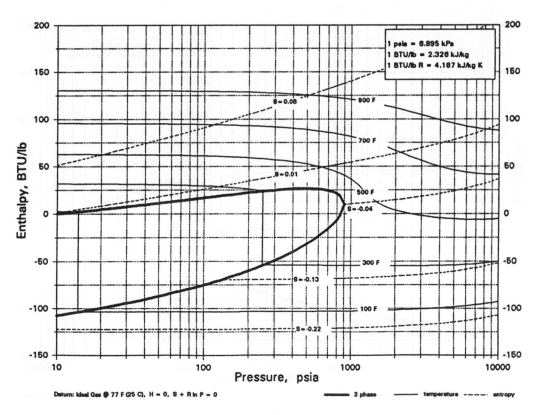

1 psia = 6.895 kPa
1 BTU/lb = 2.326 kJ/kg
1 BTU/lb R = 4.187 kJ/kg K

Datum: Ideal Gas @ 77 F (25 C), H = 0, S + R ln P = 0

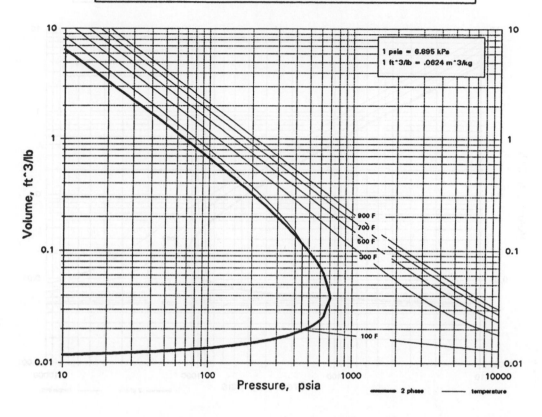

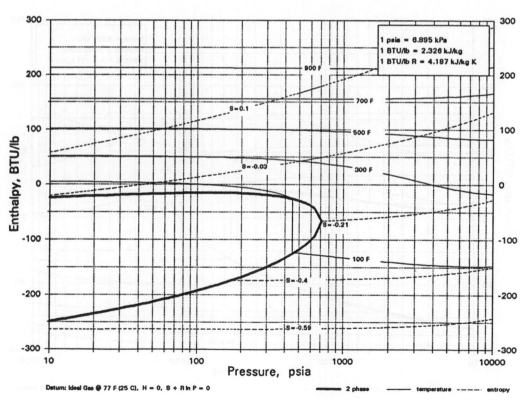

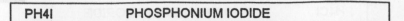

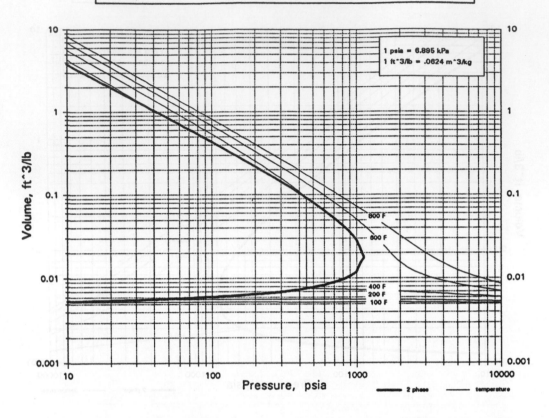

1 psia = 6.895 kPa
1 ft^3/lb = .0624 m^3/kg

2 phase   temperature

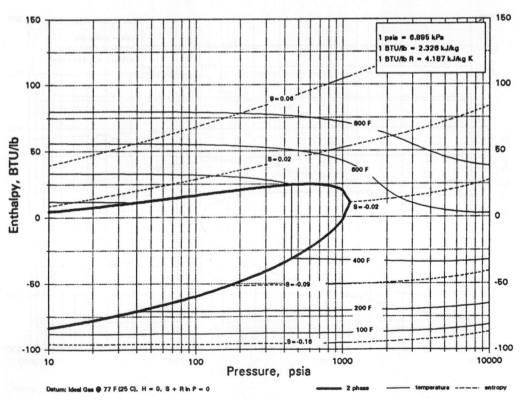

1 psia = 6.895 kPa
1 BTU/lb = 2.326 kJ/kg
1 BTU/lb R = 4.187 kJ/kg K

Datum: Ideal Gas @ 77 F (25 C), H = 0, S + R ln P = 0

2 phase   temperature   entropy

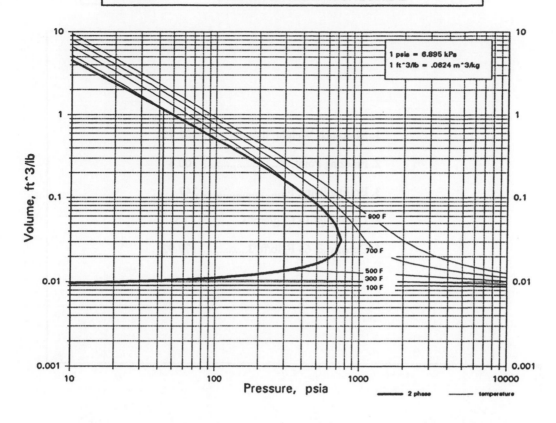

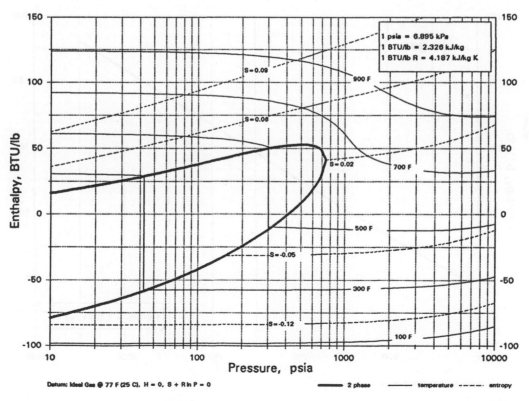

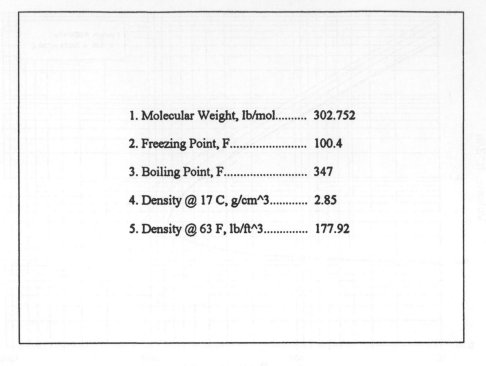

1. Molecular Weight, lb/mol.......... 302.752

2. Freezing Point, F...................... 100.4

3. Boiling Point, F........................ 347

4. Density @ 17 C, g/cm^3............ 2.85

5. Density @ 63 F, lb/ft^3.............. 177.92

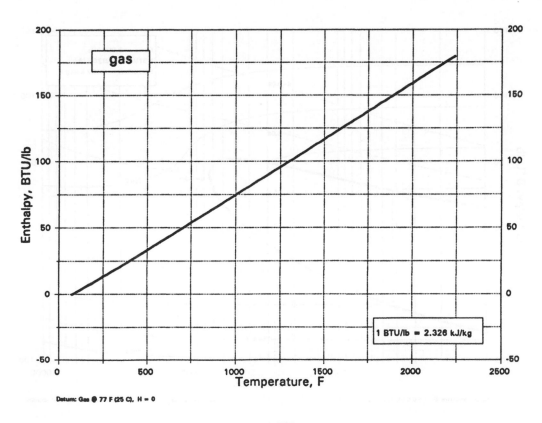

Datum: Gas @ 77 F (25 C), H = 0

222

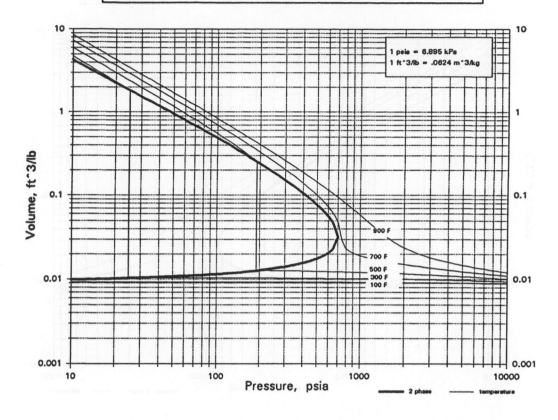

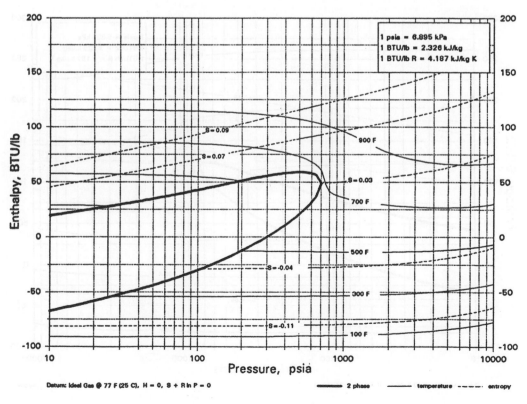

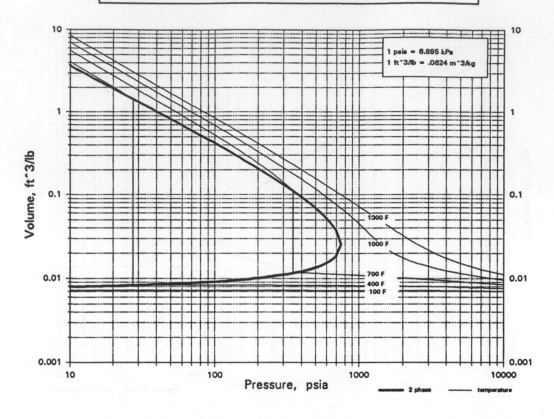

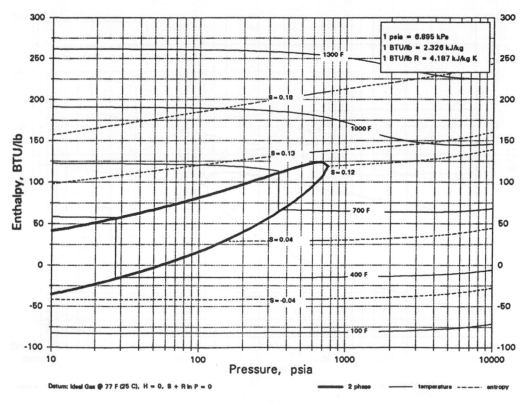

224

1. Molecular Weight, lb/mol.......... 283.889

2. Freezing Point, F....................... 788

3. Boiling Point, F.......................... ---

4. Density @ 20 C, g/cm^3............. 2.39

5. Density @ 68 F, lb/ft^3.............. 149.2

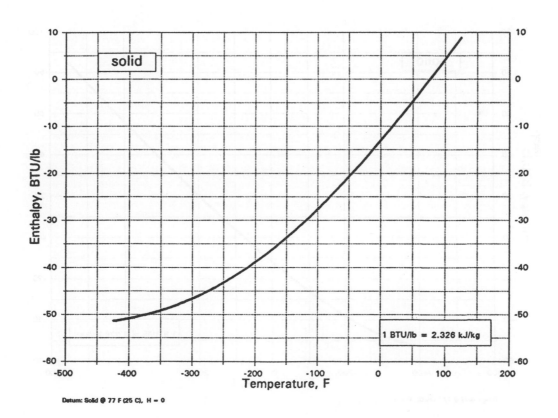

Datum: Solid @ 77 F (25 C), H = 0

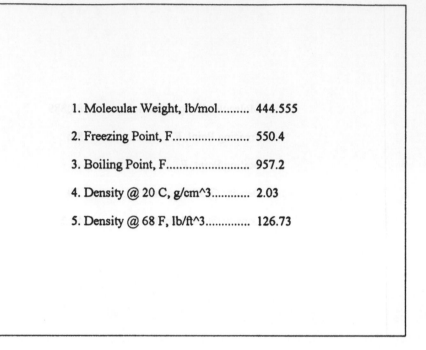

1. Molecular Weight, lb/mol.......... 444.555

2. Freezing Point, F...................... 550.4

3. Boiling Point, F......................... 957.2

4. Density @ 20 C, g/cm^3............ 2.03

5. Density @ 68 F, lb/ft^3.............. 126.73

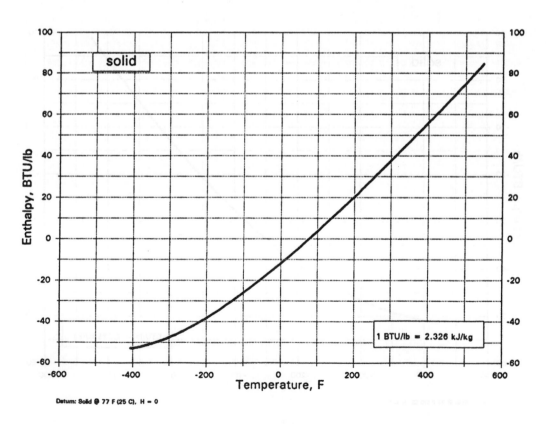

Datum: Solid @ 77 F (25 C), H = 0

226

1. Molecular Weight, lb/mol.......... 207.2

2. Freezing Point, F........................ 621.4

3. Boiling Point, F.......................... 3183.5

4. Density @ 16 C, g/cm^3............ 11.34

5. Density @ 61 F, lb/ft^3.............. 707.93

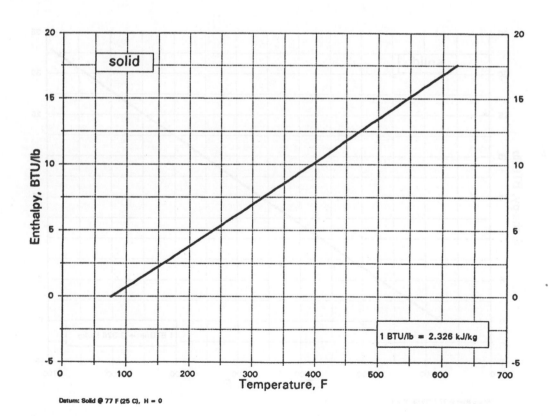

Datum: Solid @ 77 F (25 C), H = 0

227

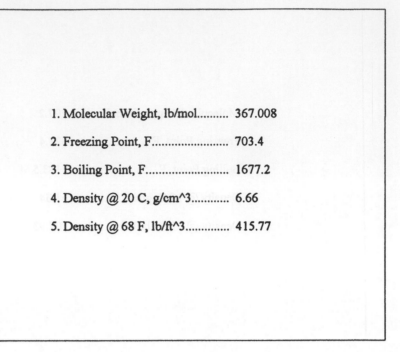

1. Molecular Weight, lb/mol.......... 367.008

2. Freezing Point, F....................... 703.4

3. Boiling Point, F......................... 1677.2

4. Density @ 20 C, g/cm^3............. 6.66

5. Density @ 68 F, lb/ft^3.............. 415.77

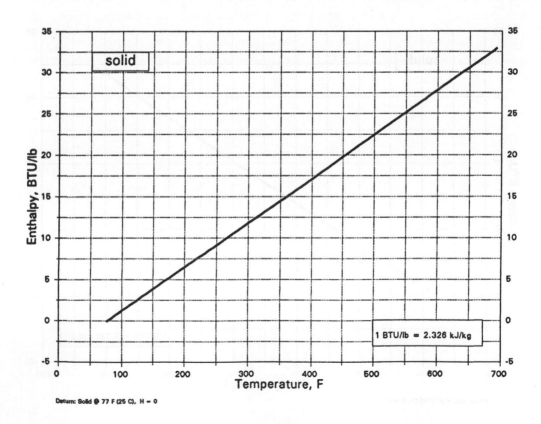

Datum: Solid @ 77 F (25 C), H = 0

228

1. Molecular Weight, lb/mol.......... 278.105

2. Freezing Point, F....................... 933.8

3. Boiling Point, F.......................... 1749.2

4. Density @ 20 C, g/cm^3............. 5.85

5. Density @ 68 F, lb/ft^3............... 365.2

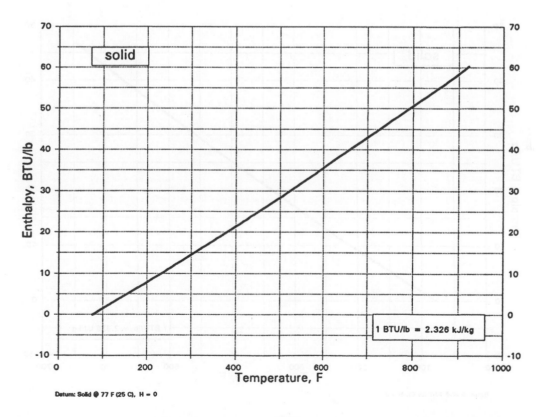

solid

1 BTU/lb = 2.326 kJ/kg

Enthalpy, BTU/lb

Temperature, F

Datum: Solid @ 77 F (25 C), H = 0

229

1. Molecular Weight, lb/mol.......... 245.197

2. Freezing Point, F........................ 1571

3. Boiling Point, F.......................... 2359.4

4. Density @ 20 C, g/cm^3............. 8.24

5. Density @ 68 F, lb/ft^3.............. 514.41

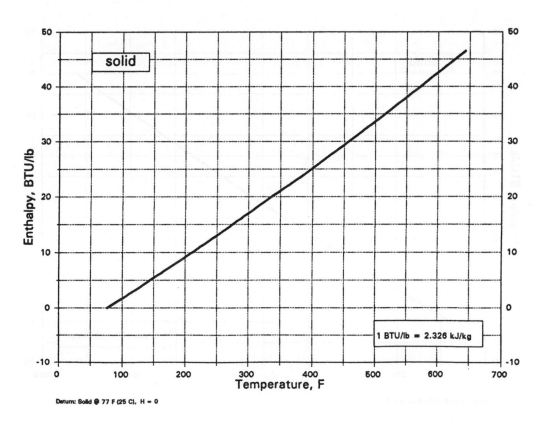

Datum: Solid @ 77 F (25 C), H = 0

PbI2	LEAD IODIDE

1. Molecular Weight, lb/mol.......... 461.009

2. Freezing Point, F........................ 755.6

3. Boiling Point, F.......................... 1601.6

4. Density @ 20 C, g/cm^3............. 6.16

5. Density @ 68 F, lb/ft^3.............. 384.56

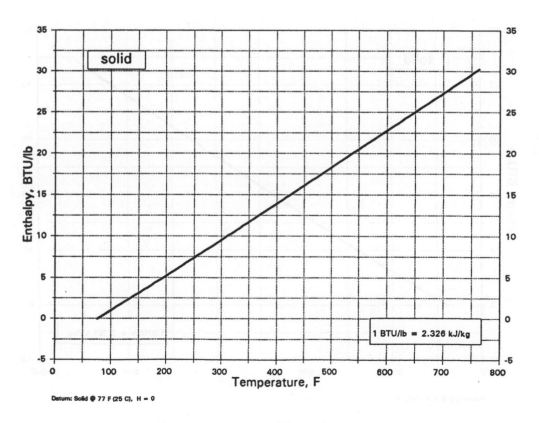

Datum: Solid @ 77 F (25 C), H = 0

231

1. Molecular Weight, lb/mol.......... 223.199

2. Freezing Point, F...................... 1634

3. Boiling Point, F......................... 2681.6

4. Density @ 20 C, g/cm^3............ 9.53

5. Density @ 68 F, lb/ft^3.............. 594.94

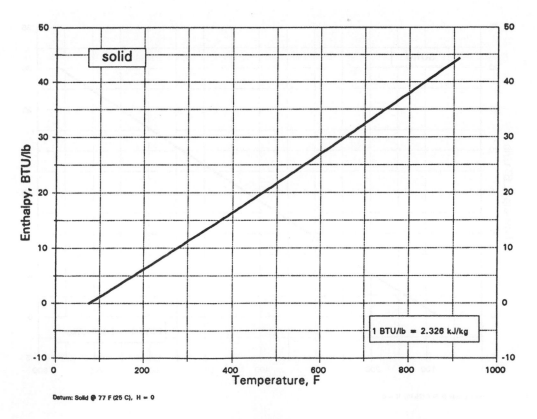

solid

1 BTU/lb = 2.326 kJ/kg

Datum: Solid @ 77 F (25 C), H = 0

1. Molecular Weight, lb/mol.......... 239.266

2. Freezing Point, F...................... 2037.2

3. Boiling Point, F......................... 2337.8

4. Density @ 20 C, g/cm^3............ 7.5

5. Density @ 68 F, lb/ft^3.............. 468.21

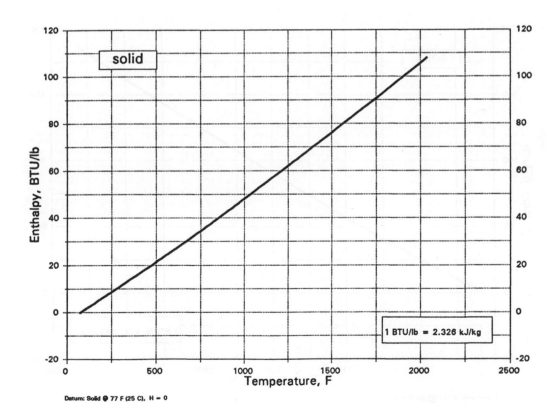

Datum: Solid @ 77 F (25 C), H = 0

1. Molecular Weight, lb/mol.......... 106.42

2. Freezing Point, F........................ 2830.8

3. Boiling Point, F........................... 5633.3

4. Density @ 20 C, g/cm^3............ 12.02

5. Density @ 68 F, lb/ft^3.............. 750.38

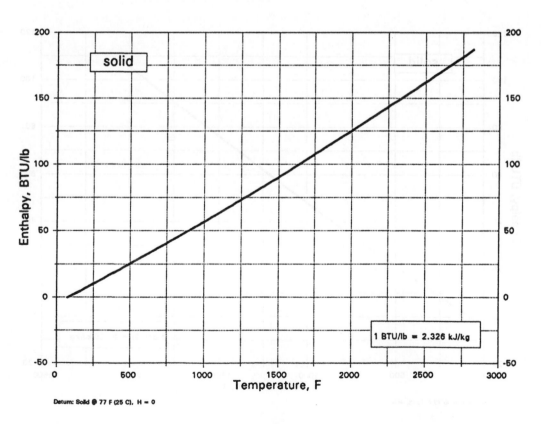

Datum: Solid @ 77 F (25 C), H = 0

1. Molecular Weight, lb/mol.......... 209

2. Freezing Point, F........................ 489.2

3. Boiling Point, F.......................... 1763.3

4. Density @ 20 C, g/cm^3............ 9.4

5. Density @ 68 F, lb/ft^3.............. 586.82

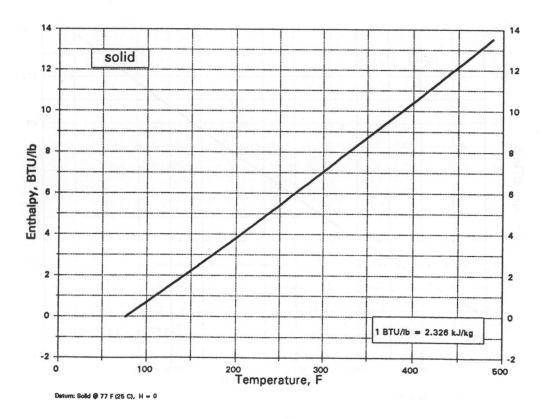

solid

1 BTU/lb = 2.326 kJ/kg

Datum: Solid @ 77 F (25 C), H = 0

235

1. Molecular Weight, lb/mol.......... 195.08

2. Freezing Point, F........................ 3215.1

3. Boiling Point, F......................... 6704.3

4. Density @ 20 C, g/cm^3............ 21.45

5. Density @ 68 F, lb/ft^3.............. 1339.08

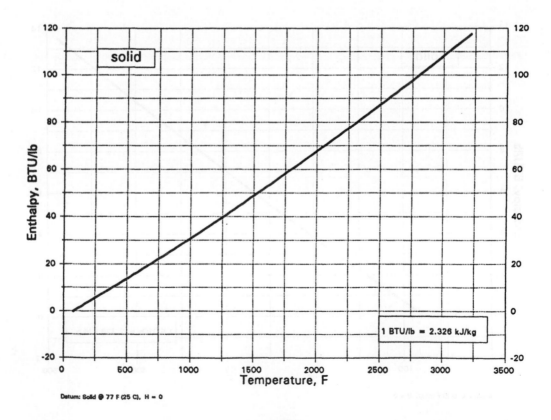

Datum: Solid @ 77 F (25 C), H = 0

1. Molecular Weight, lb/mol.......... 226

2. Freezing Point, F....................... 1292

3. Boiling Point, F.......................... 2796.5

4. Density @ 20 C, g/cm^3............ 5

5. Density @ 68 F, lb/ft^3.............. 312.14

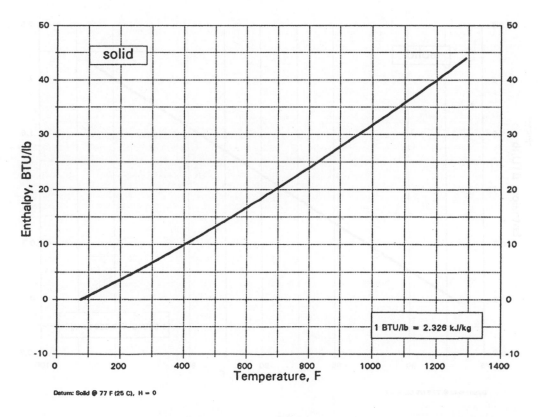

solid

Enthalpy, BTU/lb

Temperature, F

1 BTU/lb = 2.326 kJ/kg

Datum: Solid @ 77 F (25 C), H = 0

237

1. Molecular Weight, lb/mol.......... 85.468

2. Freezing Point, F........................ 102.7

3. Boiling Point, F.......................... 1300.7

4. Density @ 20 C, g/cm^3............. 1.53

5. Density @ 68 F, lb/ft^3.............. 95.51

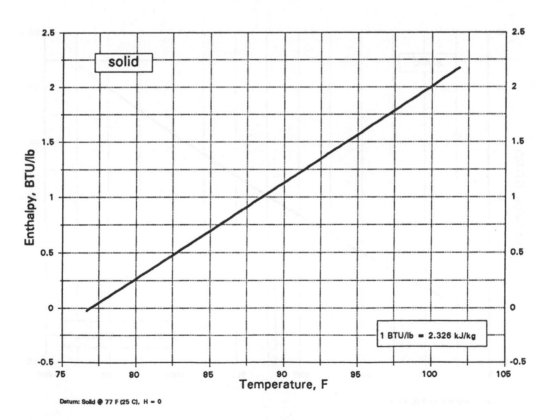

solid

1 BTU/lb = 2.326 kJ/kg

Enthalpy, BTU/lb

Temperature, F

Datum: Solid @ 77 F (25 C), H = 0

238

1. Molecular Weight, lb/mol.......... 165.372

2. Freezing Point, F........................ 1259.6

3. Boiling Point, F.......................... 2465.6

4. Density @ 20 C, g/cm^3............ 3.35

5. Density @ 68 F, lb/ft^3.............. 209.13

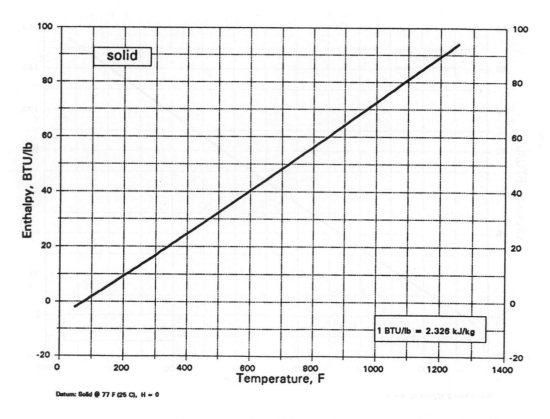

Datum: Solid @ 77 F (25 C), H = 0

1 BTU/lb = 2.326 kJ/kg

1. Molecular Weight, lb/mol.......... 120.921

2. Freezing Point, F....................... 1319

3. Boiling Point, F.......................... 2517.8

4. Density @ 20 C, g/cm^3............ 2.8

5. Density @ 68 F, lb/ft^3.............. 174.8

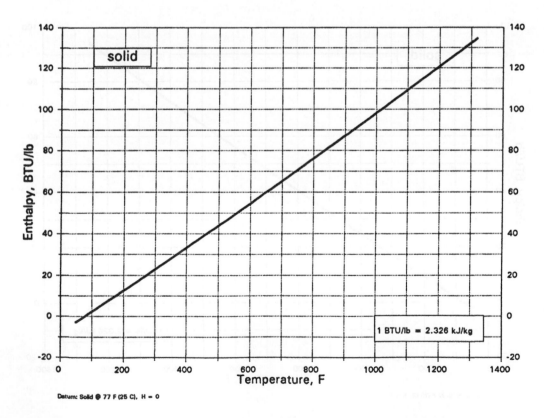

Datum: Solid @ 77 F (25 C), H = 0

1. Molecular Weight, lb/mol.......... 104.466

2. Freezing Point, F........................ 1400

3. Boiling Point, F.......................... 2566.4

4. Density @ 20 C, g/cm^3............. 3.56

5. Density @ 68 F, lb/ft^3.............. 222.24

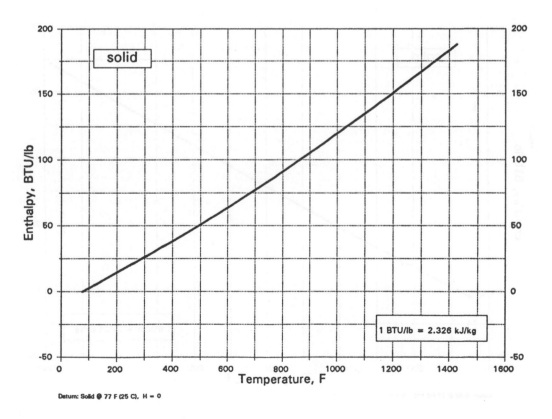

solid

1 BTU/lb = 2.326 kJ/kg

Datum: Solid @ 77 F (25 C), H = 0

1. Molecular Weight, lb/mol.......... 212.372

2. Freezing Point, F...................... 1187.6

3. Boiling Point, F......................... 2379.2

4. Density @ 20 C, g/cm^3............ 3.55

5. Density @ 68 F, lb/ft^3.............. 221.62

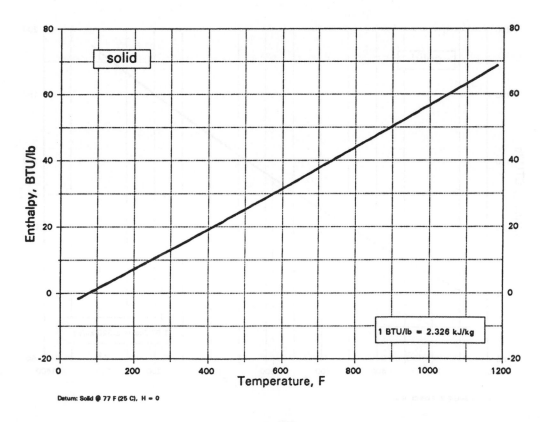

Datum: Solid @ 77 F (25 C), H = 0

1. Molecular Weight, lb/mol.......... 186.207

2. Freezing Point, F........................ 5766.8

3. Boiling Point, F.......................... 10187.3

4. Density @ 20 C, g/cm^3............ 20.53

5. Density @ 68 F, lb/ft^3.............. 1281.64

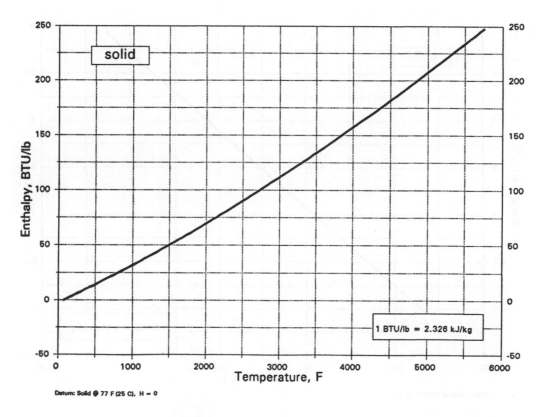

Datum: Solid @ 77 F (25 C), H = 0

1. Molecular Weight, lb/mol.......... 484.41

2. Freezing Point, F...................... 564.8

3. Boiling Point, F......................... 684.3

4. Density @ 20 C, g/cm^3............ 6.1

5. Density @ 68 F, lb/ft^3.............. 380.81

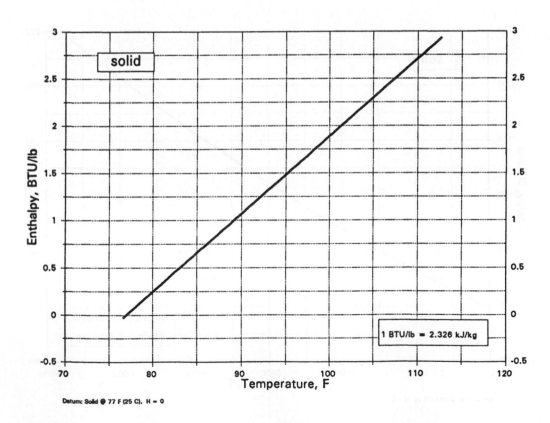

Datum: Solid @ 77 F (25 C). H = 0

1. Molecular Weight, lb/mol.......... 102.906

2. Freezing Point, F........................ 3567.2

3. Boiling Point, F......................... 6632.3

4. Density @ 20 C, g/cm^3............. 12.4

5. Density @ 68 F, lb/ft^3.............. 774.1

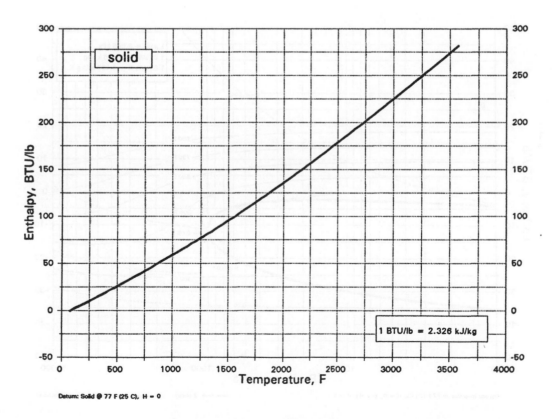

solid

1 BTU/lb = 2.326 kJ/kg

Datum: Solid @ 77 F (25 C), H = 0

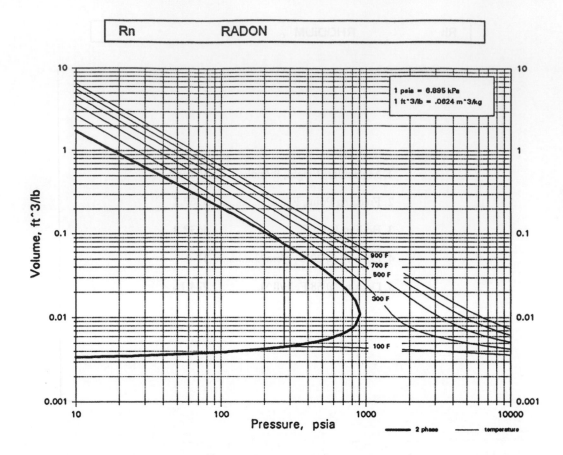

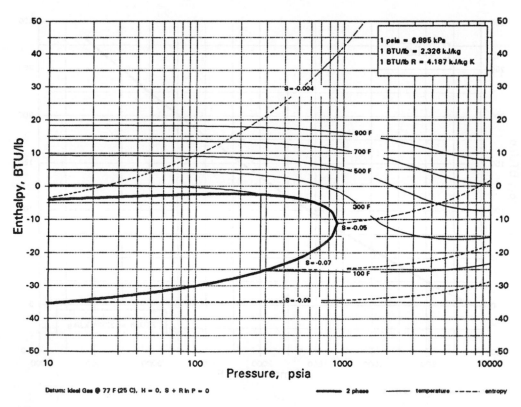

246

1. Molecular Weight, lb/mol.......... 101.07

2. Freezing Point, F........................ 4233.2

3. Boiling Point, F........................ 7640.3

4. Density @ 20 C, g/cm^3............ 12.3

5. Density @ 68 F, lb/ft^3.............. 767.86

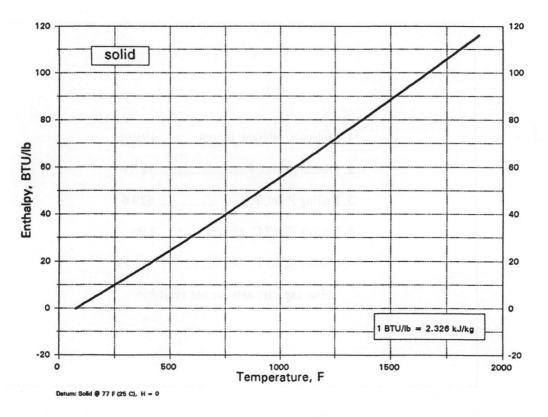

Datum: Solid @ 77 F (25 C), H = 0

1. Molecular Weight, lb/mol.......... 196.062

2. Freezing Point, F....................... 187.7

3. Boiling Point, F......................... 620.6

4. Density @ 17 C, g/cm^3............ 2.96

5. Density @ 63 F, lb/ft^3.............. 184.79

---

1. Molecular Weight, lb/mol.......... 196.062

2. Freezing Point, F....................... 187.7

3. Boiling Point, F......................... 620.6

4. Density @ 17 C, g/cm^3............ 2.96

5. Density @ 63 F, lb/ft^3.............. 184.79

Heat capacity data are not available.

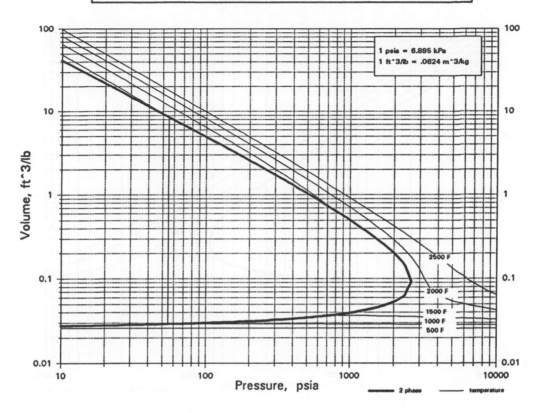

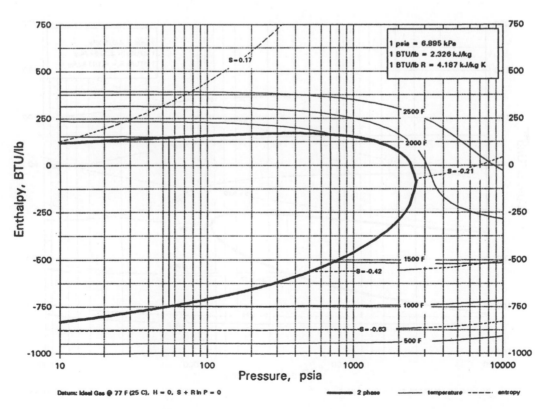

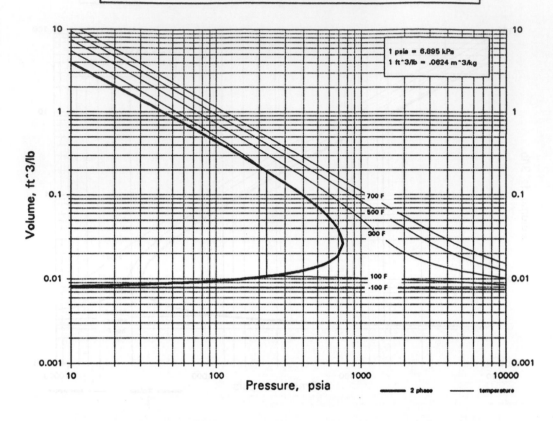

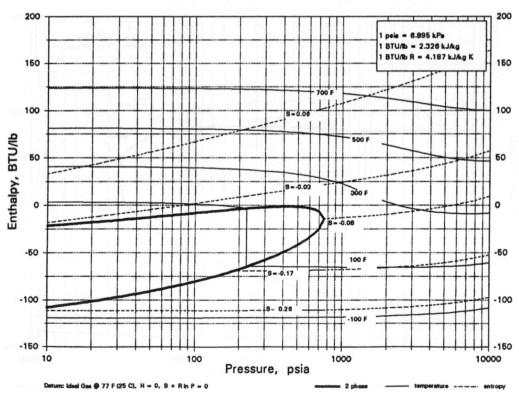

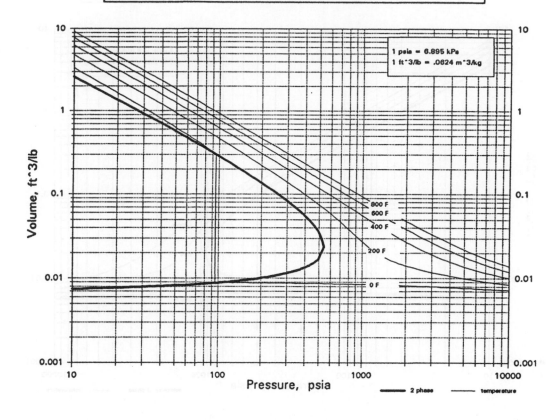

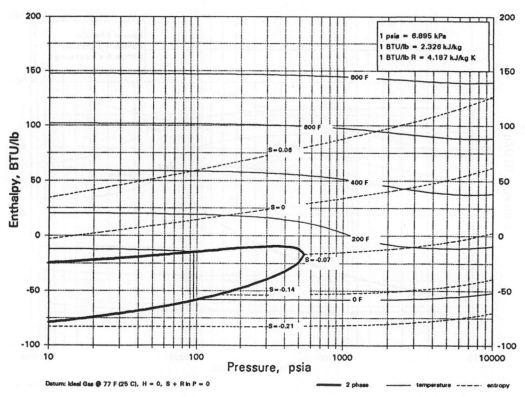

251

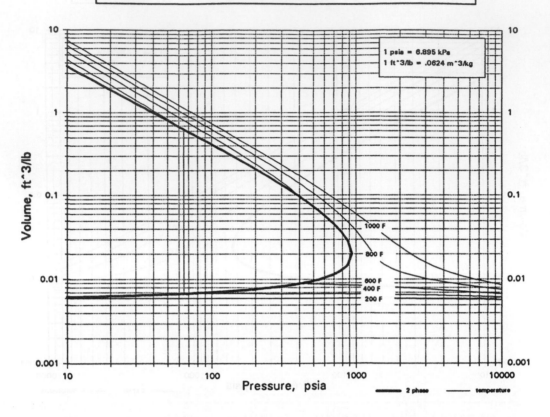

1 psia = 6.895 kPa
1 ft^3/lb = .0624 m^3/kg

1000 F
800 F
600 F
400 F
200 F

2 phase   temperature

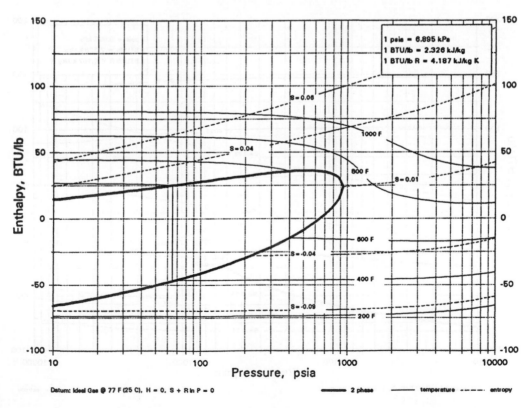

1 psia = 6.895 kPa
1 BTU/lb = 2.326 kJ/kg
1 BTU/lb R = 4.187 kJ/kg K

S = 0.06
S = 0.04
S = 0.01
1000 F
800 F
600 F
S = -0.04
400 F
S = -0.09
200 F

Datum: Ideal Gas @ 77 F (25 C), H = 0, S + R ln P = 0

2 phase   temperature   entropy

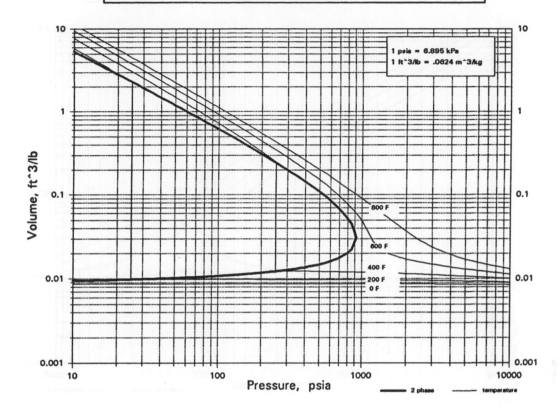

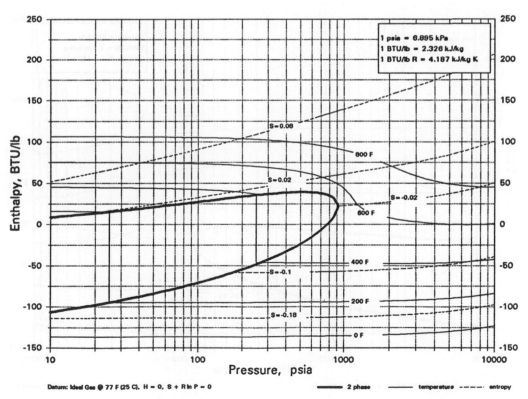

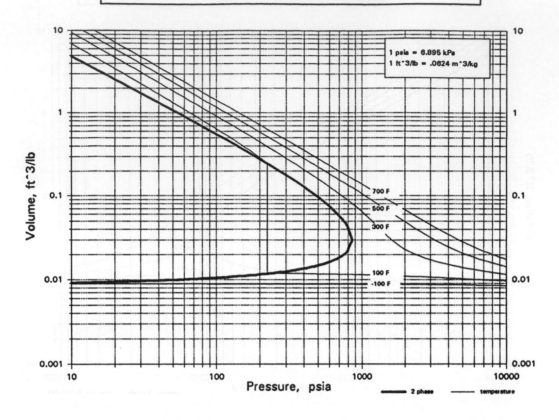

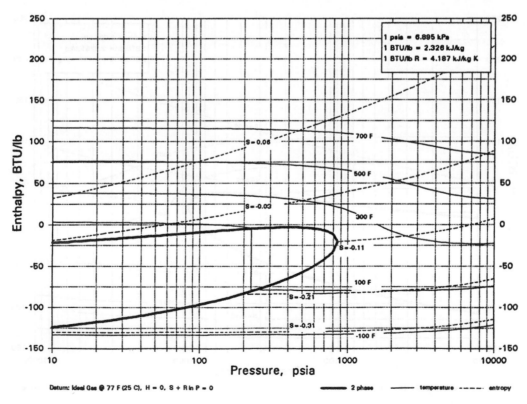

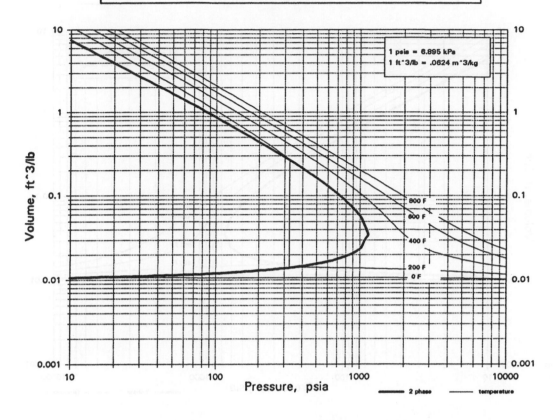

1 psia = 6.895 kPa
1 ft^3/lb = .0624 m^3/kg

800 F
600 F
400 F
200 F
0 F

Volume, ft^3/lb

Pressure, psia

2 phase   temperature

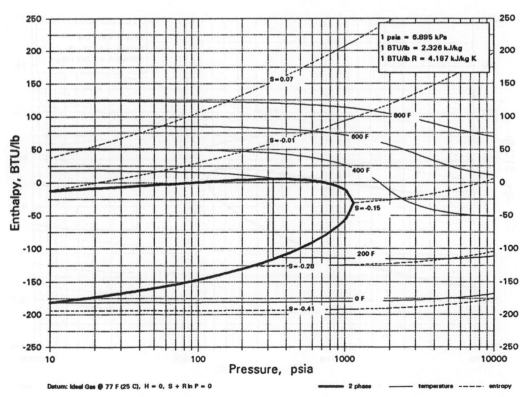

1 psia = 6.895 kPa
1 BTU/lb = 2.326 kJ/kg
1 BTU/lb R = 4.187 kJ/kg K

S=0.07
S=-0.01
S=-0.15
S=-0.26
S=-0.41

800 F
600 F
400 F
200 F
0 F

Enthalpy, BTU/lb

Pressure, psia

Datum: Ideal Gas @ 77 F (25 C), H = 0, S + R ln P = 0

2 phase   temperature   entropy

255

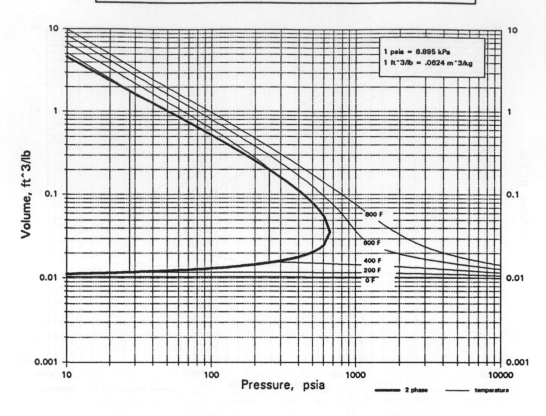

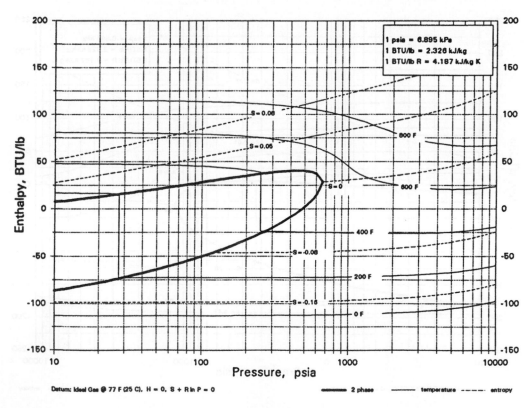

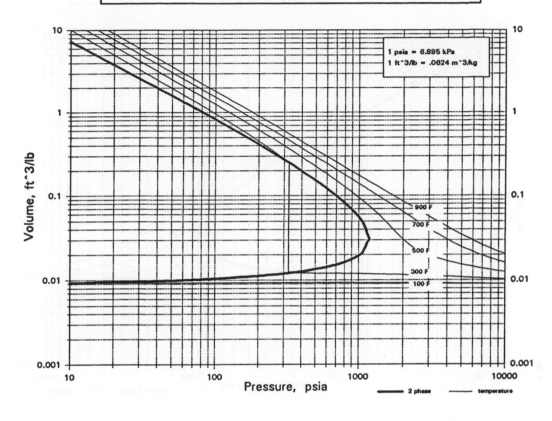

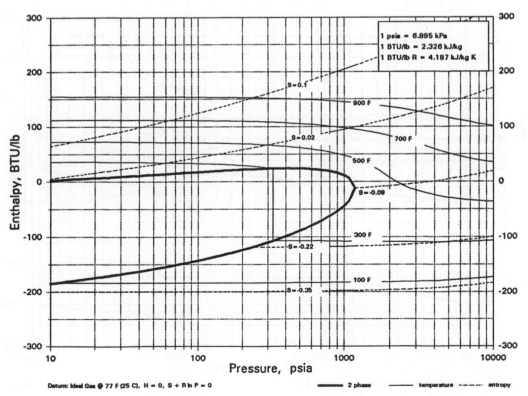

257

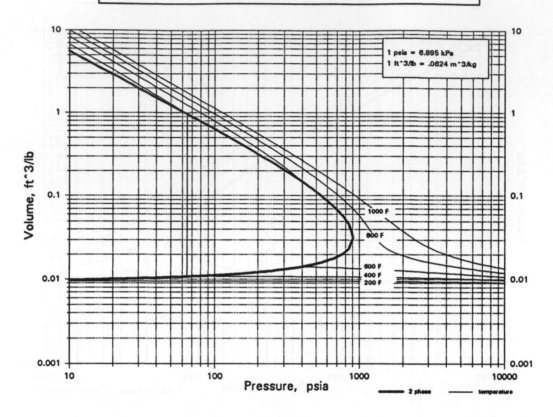

1 psia = 6.895 kPa
1 ft^3/lb = .0624 m^3/kg

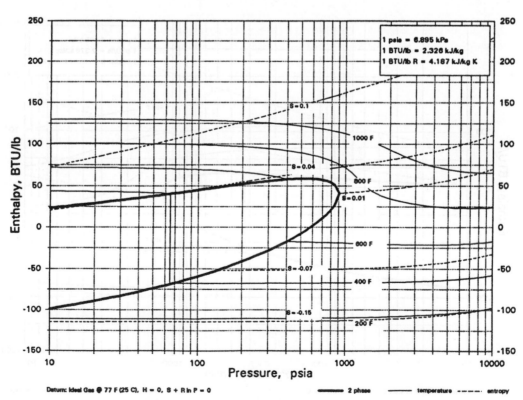

1 psia = 6.895 kPa
1 BTU/lb = 2.326 kJ/kg
1 BTU/lb R = 4.187 kJ/kg K

Datum: Ideal Gas @ 77 F (25 C), H = 0, S + R ln P = 0

1. Molecular Weight, lb/mol.......... 121.757

2. Freezing Point, F........................ 1167.1

3. Boiling Point, F.......................... 2956.7

4. Density @ 25 C, g/cm^3............. 6.68

5. Density @ 77 F, lb/ft^3.............. 417.02

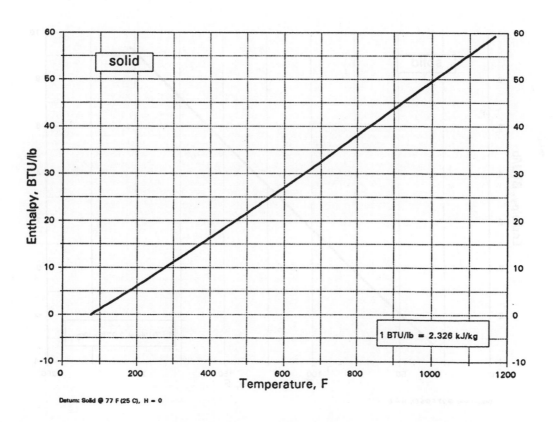

solid

1 BTU/lb = 2.326 kJ/kg

Enthalpy, BTU/lb

Temperature, F

Datum: Solid @ 77 F (25 C), H = 0

1. Molecular Weight, lb/mol.......... 361.469

2. Freezing Point, F........................ 205.8

3. Boiling Point, F.......................... 527

4. Density @ 23 C, g/cm^3............ 4.15

5. Density @ 73 F, lb/ft^3.............. 259.08

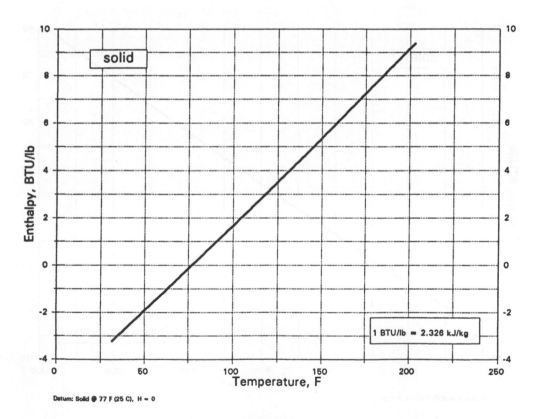

Datum: Solid @ 77 F (25 C), H = 0

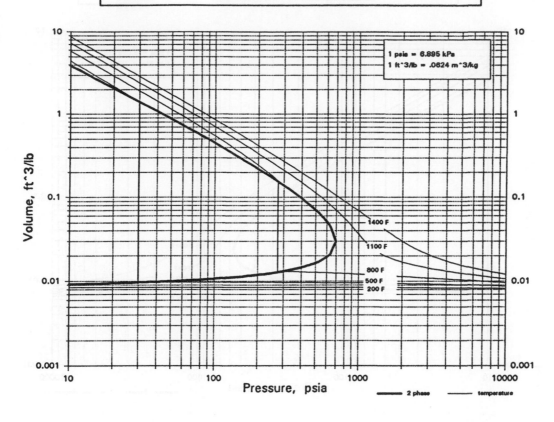

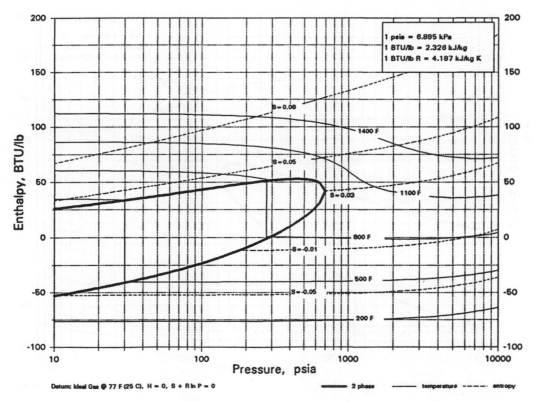

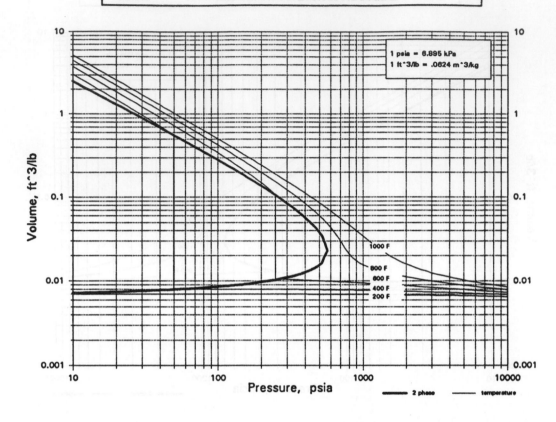

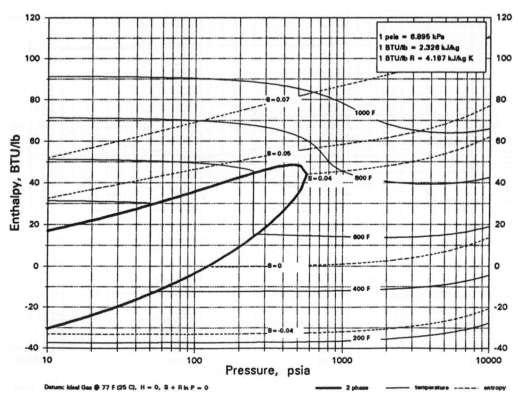

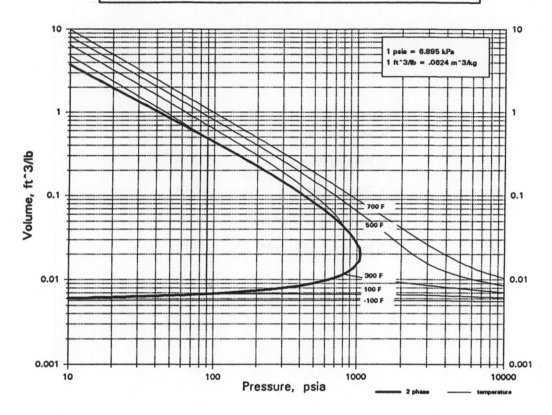

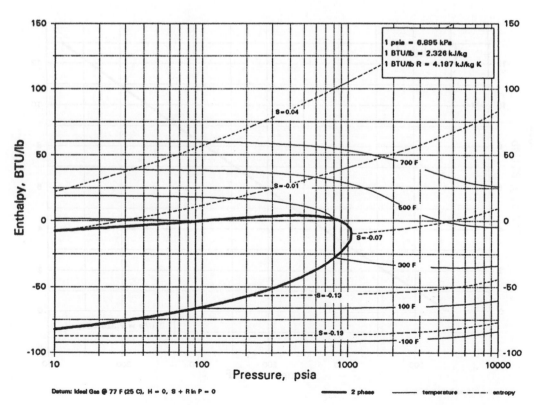

1. Molecular Weight, lb/mol.......... 502.47

2. Freezing Point, F...................... 332.6

3. Boiling Point, F........................ 753.8

4. Density @ 17 C, g/cm^3............ 4.92

5. Density @ 63 F, lb/ft^3.............. 307.14

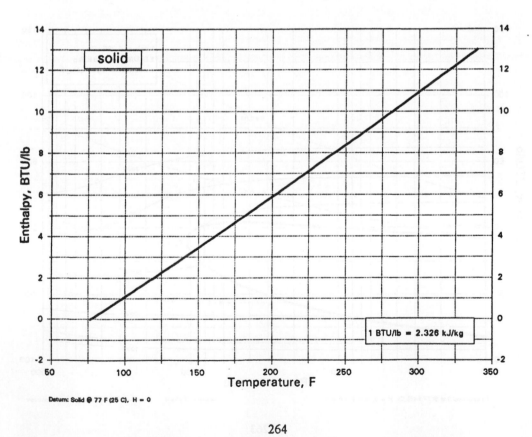

Datum: Solid @ 77 F (25 C), H = 0

1. Molecular Weight, lb/mol.......... 291.512

2. Freezing Point, F........................ 1212.8

3. Boiling Point, F.......................... 2597

4. Density @ 20 C, g/cm^3............. 5.2

5. Density @ 68 F, lb/ft^3.............. 324.62

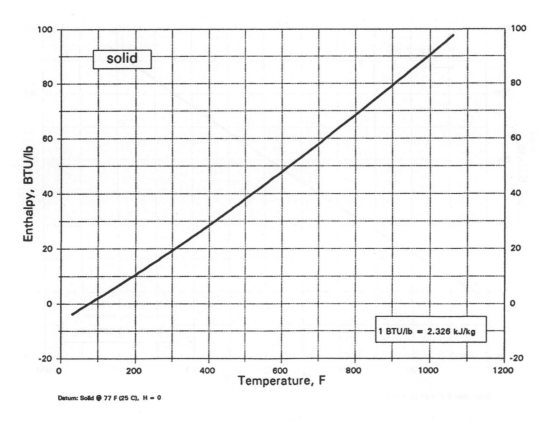

Datum: Solid @ 77 F (25 C), H = 0

1. Molecular Weight, lb/mol.......... 44.956

2. Freezing Point, F........................ 2805.8

3. Boiling Point, F.......................... 4400.3

4. Density @ 20 C, g/cm^3............ 2.99

5. Density @ 68 F, lb/ft^3.............. 186.66

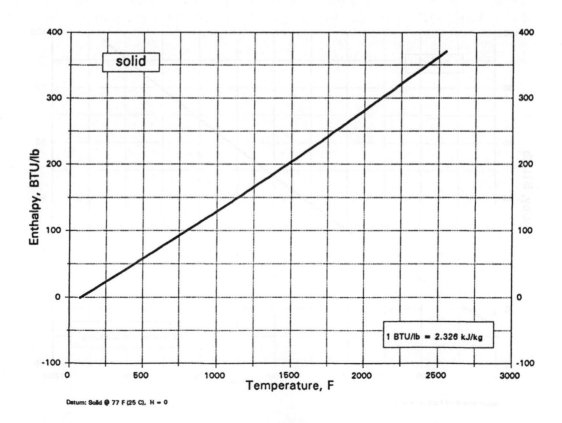

Datum: Solid @ 77 F (25 C), H = 0

1. Molecular Weight, lb/mol.......... 78.96

2. Freezing Point, F........................ 429.8

3. Boiling Point, F......................... 1214.3

4. Density @ 20 C, g/cm^3............ 4.81

5. Density @ 68 F, lb/ft^3.............. 300.28

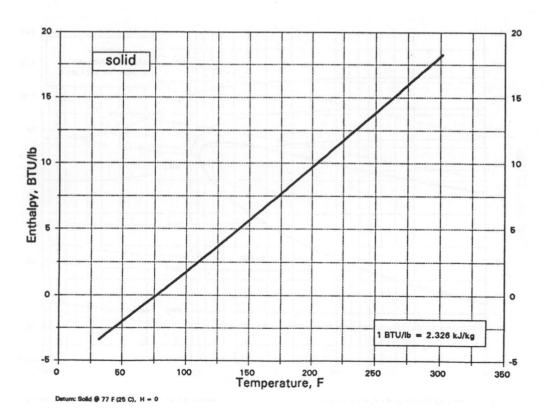

Datum: Solid @ 77 F (25 C), H = 0

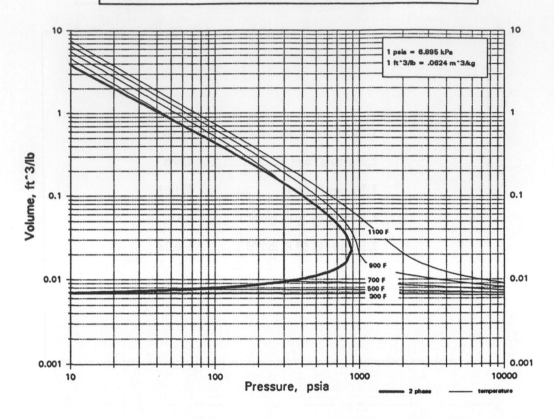

1 psia = 6.895 kPa
1 ft^3/lb = .0624 m^3/kg

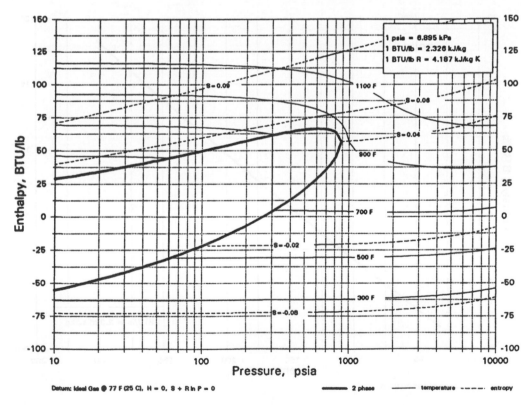

1 psia = 6.895 kPa
1 BTU/lb = 2.326 kJ/kg
1 BTU/lb R = 4.187 kJ/kg K

Datum: Ideal Gas @ 77 F (25 C), H = 0, S + R ln P = 0

268

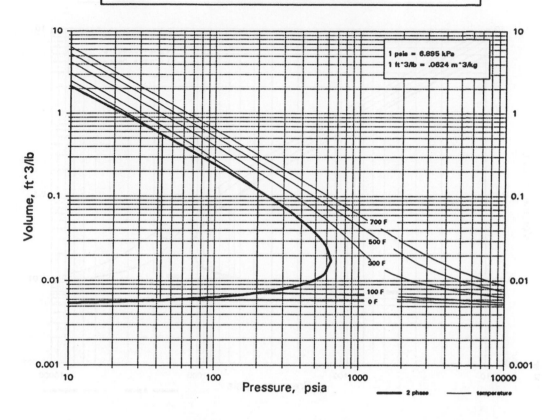

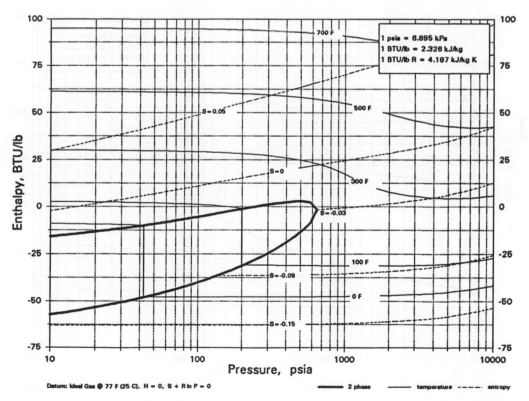

Datum: Ideal Gas @ 77 F (25 C), H = 0, S + R ln P = 0

SeOCl2   SELENIUM OXYCHLORIDE

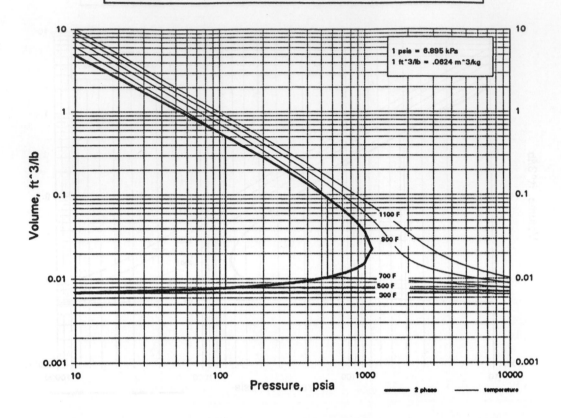

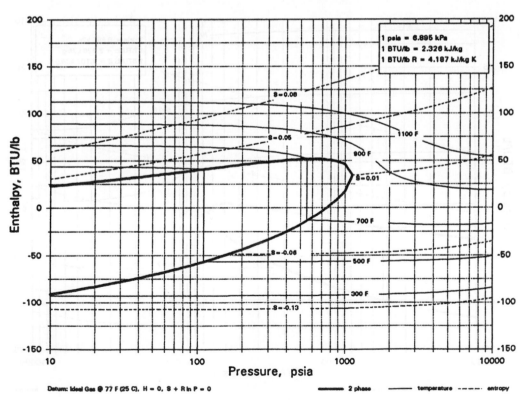

Datum: Ideal Gas @ 77 F (25 C), H = 0, S + R ln P = 0

270

1. Molecular Weight, lb/mol.......... 110.959

2. Freezing Point, F........................ 644

3. Boiling Point, F........................... 602.6

4. Density @ 15 C, g/cm^3............. 3.95

5. Density @ 59 F, lb/ft^3.............. 246.59

---

1. Molecular Weight, lb/mol.......... 110.959

2. Freezing Point, F........................ 644

3. Boiling Point, F........................... 602.6

4. Density @ 15 C, g/cm^3............. 3.95

5. Density @ 59 F, lb/ft^3.............. 246.59

Heat capacity data are not available.

1. Molecular Weight, lb/mol........... 28.086

2. Freezing Point, F...................... 2573.3

3. Boiling Point, F......................... 5865.1

4. Density @ 20 C, g/cm^3............. 2.34

5. Density @ 68 F, lb/ft^3.............. 146.08

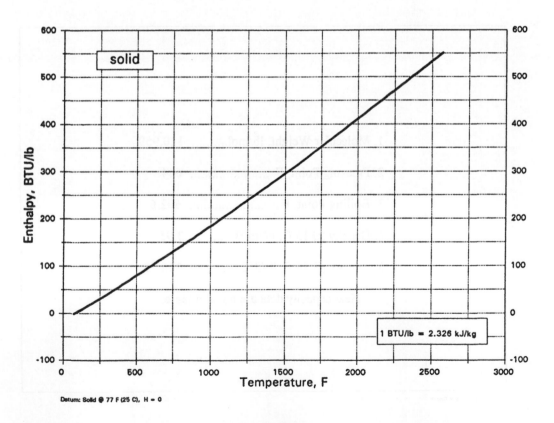

Datum: Solid @ 77 F (25 C), H = 0

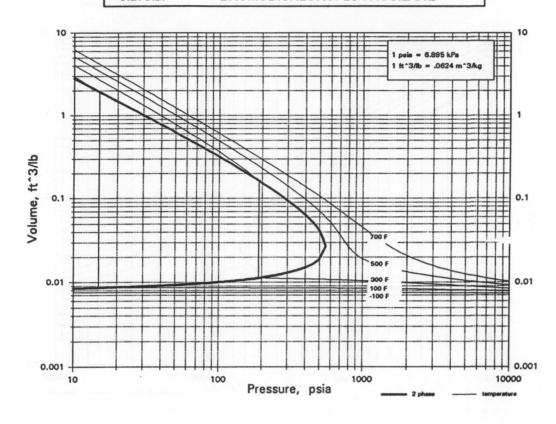

1 psia = 6.895 kPa
1 ft^3/lb = .0624 m^3/kg

Volume, ft^3/lb

700 F
500 F
300 F
100 F
-100 F

Pressure, psia

2 phase   temperature

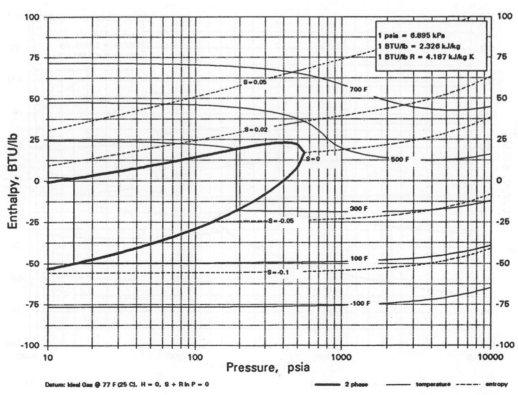

1 psia = 6.895 kPa
1 BTU/lb = 2.326 kJ/kg
1 BTU/lb R = 4.187 kJ/kg K

Enthalpy, BTU/lb

S=0.05
S=0.02
700 F
S=0
500 F
300 F
S=-0.05
100 F
S=-0.1
-100 F

Pressure, psia

Datum: Ideal Gas @ 77 F (25 C). H = 0, S + R ln P = 0   2 phase   temperature   entropy

273

SiBrF3   TRIFLUOROBROMOSILANE

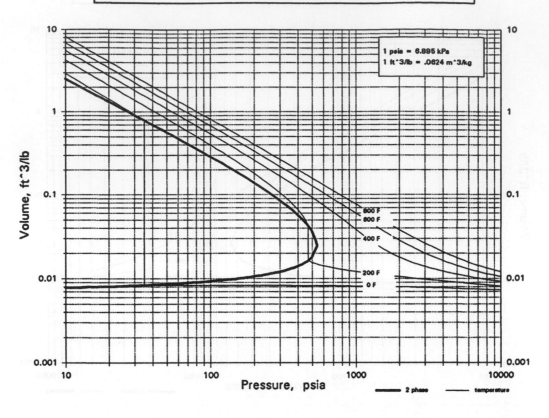

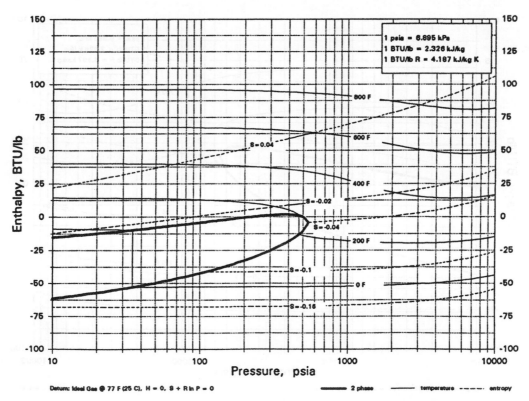

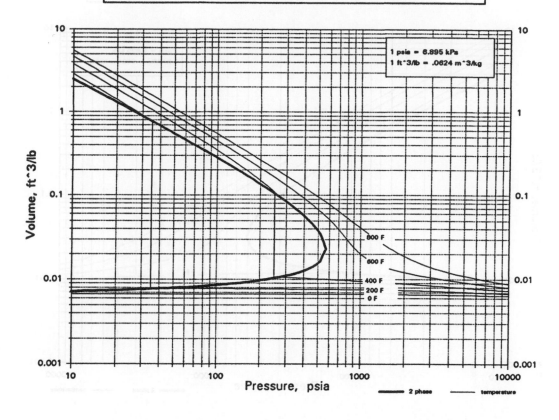

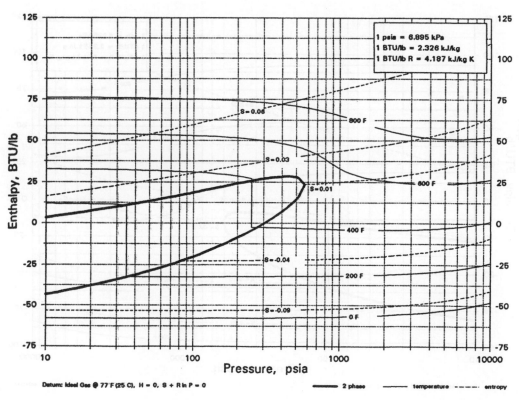

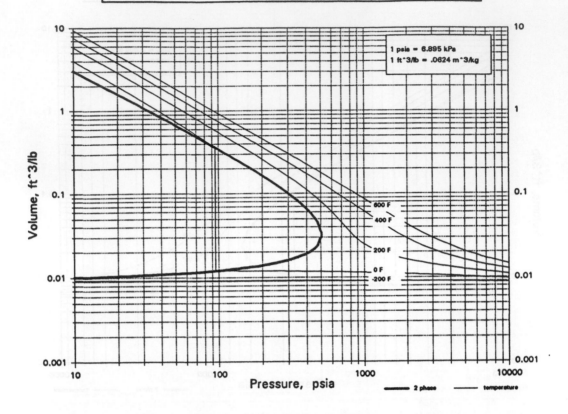

1 psia = 6.895 kPa
1 ft^3/lb = .0624 m^3/kg

600 F
400 F
200 F
0 F
-200 F

Pressure, psia

━━━ 2 phase   ─── temperature

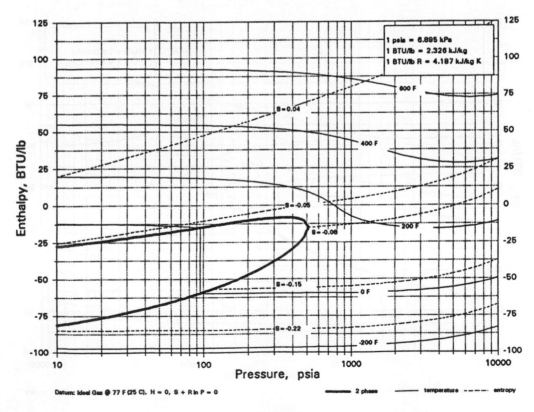

1 psia = 6.895 kPa
1 BTU/lb = 2.326 kJ/kg
1 BTU/lb R = 4.187 kJ/kg K

S=0.04
600 F
400 F
S=-0.05
S=-0.06
200 F
S=-0.15
0 F
S=-0.22
-200 F

Pressure, psia

Datum: Ideal Gas @ 77 F (25 C), H = 0, S + R ln P = 0   ━━━ 2 phase   ─── temperature   ----- entropy

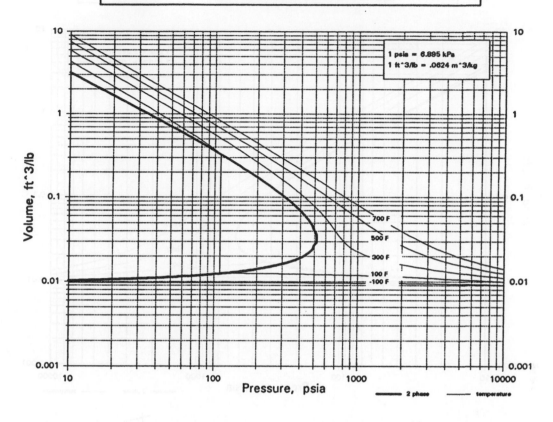

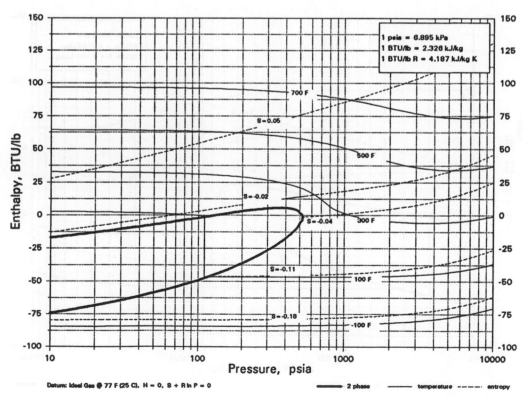

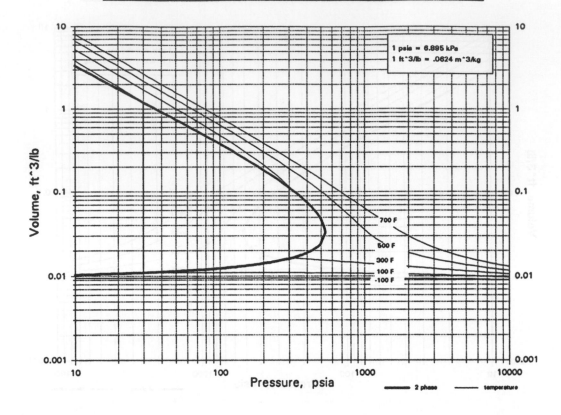

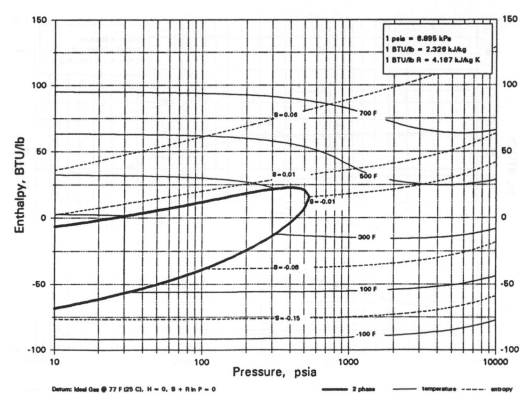

278

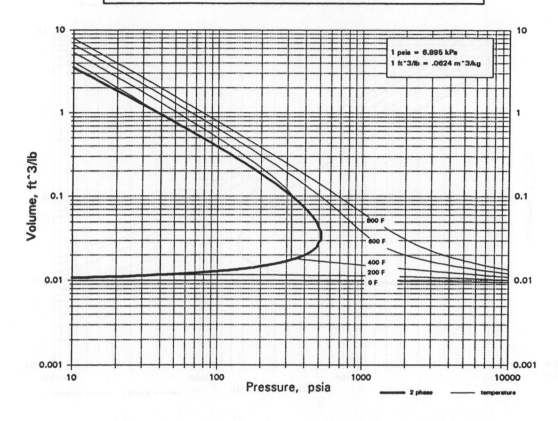

1 psia = 6.895 kPa
1 ft^3/lb = .0624 m^3/kg

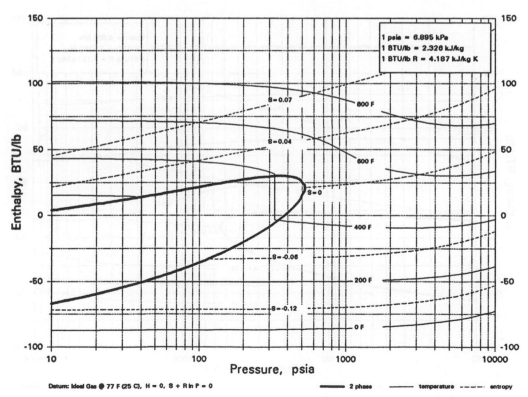

1 psia = 6.895 kPa
1 BTU/lb = 2.326 kJ/kg
1 BTU/lb R = 4.187 kJ/kg K

Datum: Ideal Gas @ 77 F (25 C), H = 0, S + R ln P = 0

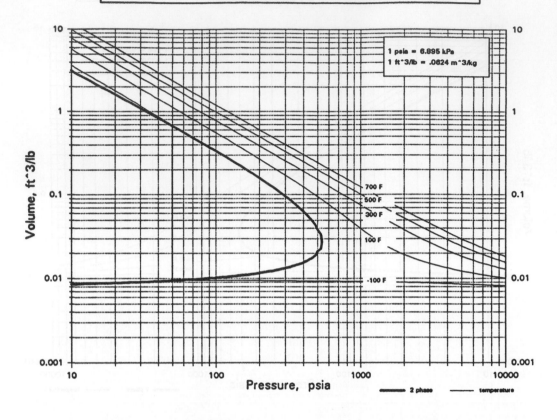

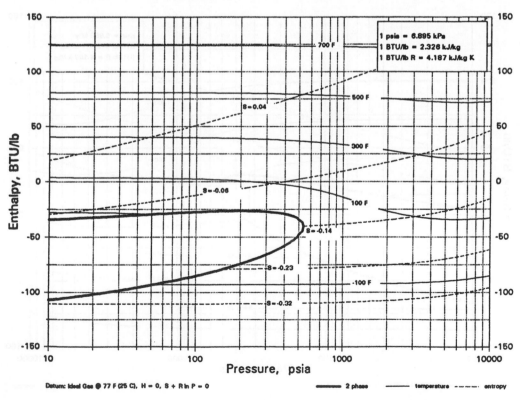

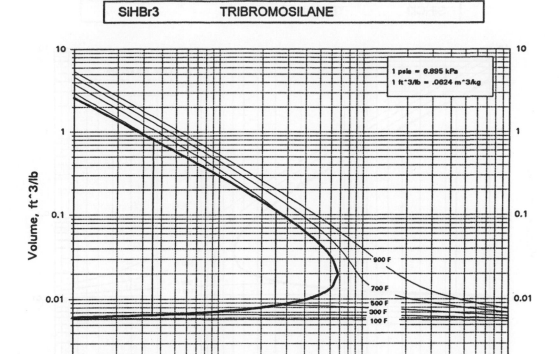

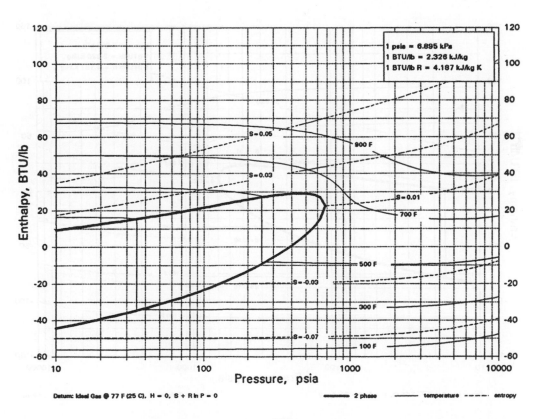

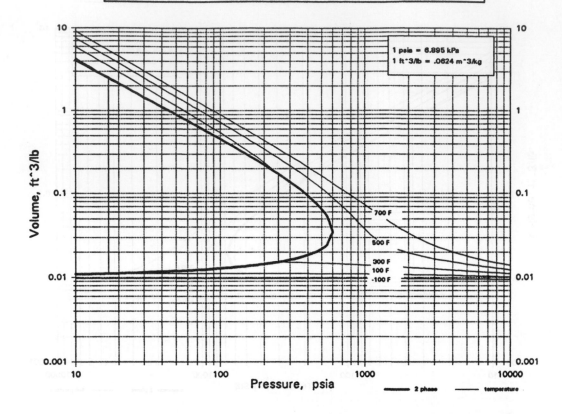

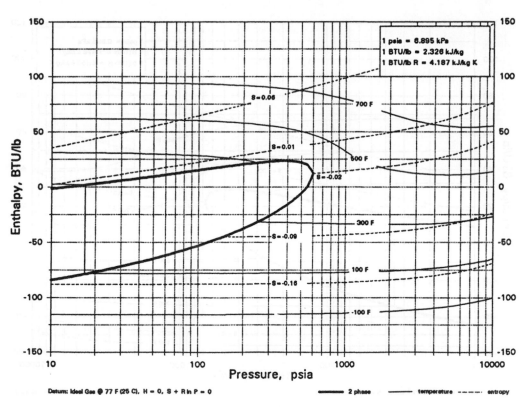

282

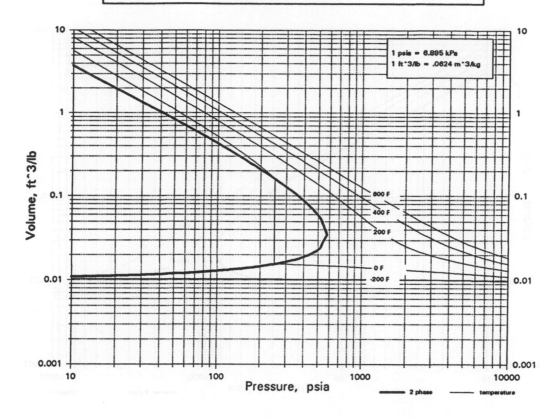

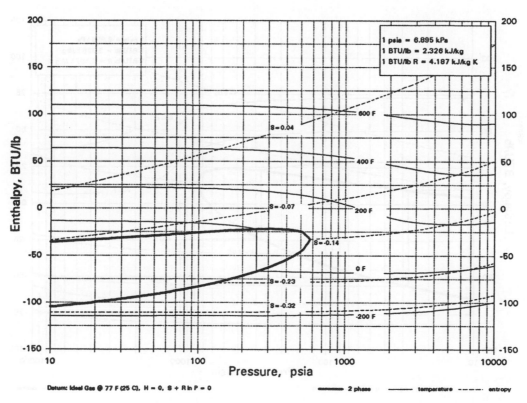

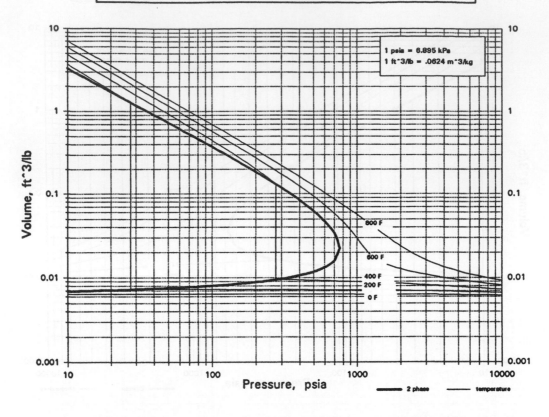

1 psia = 6.895 kPa
1 ft^3/lb = .0624 m^3/kg

——— 2 phase   ——— temperature

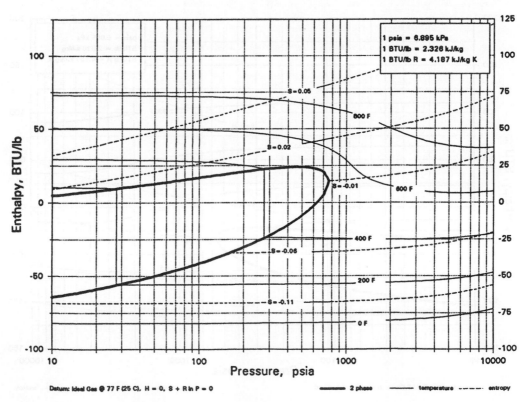

1 psia = 6.895 kPa
1 BTU/lb = 2.326 kJ/kg
1 BTU/lb R = 4.187 kJ/kg K

Datum: Ideal Gas @ 77 F (25 C), H = 0, S + R ln P = 0

——— 2 phase   ——— temperature   - - - - - entropy

284

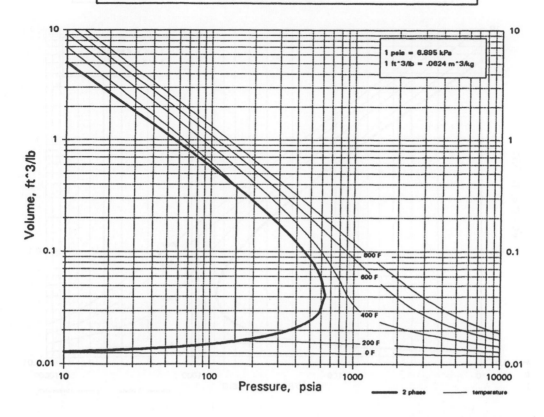

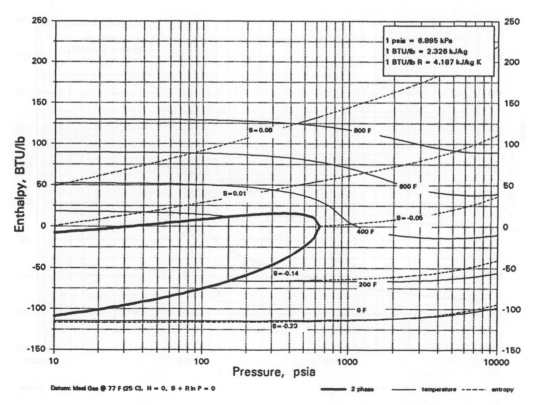

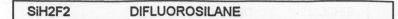

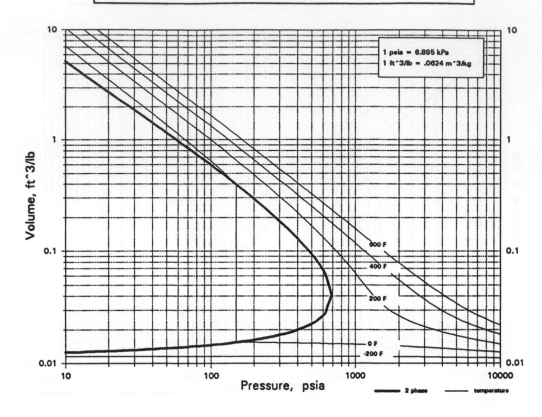

1 psia = 6.895 kPa
1 ft^3/lb = .0624 m^3/kg

600 F
400 F
200 F
0 F
-200 F

2 phase   temperature

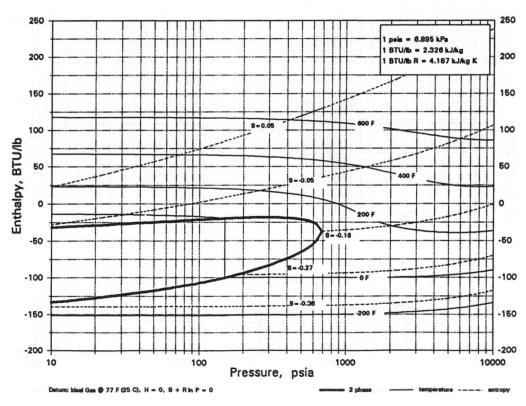

1 psia = 6.895 kPa
1 BTU/lb = 2.326 kJ/kg
1 BTU/lb R = 4.187 kJ/kg K

S = 0.05
S = -0.05
S = -0.16
S = -0.27
S = -0.36

600 F
400 F
200 F
0 F
-200 F

Datum: Ideal Gas @ 77 F (25 C), H = 0, S + R ln P = 0

2 phase   temperature   entropy

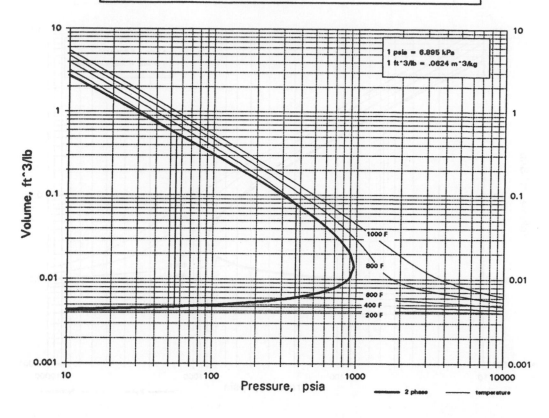

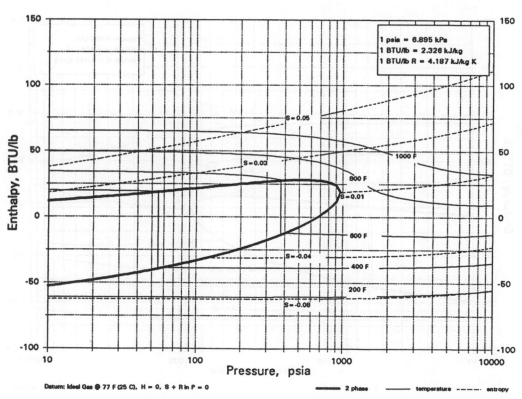

Datum: Ideal Gas @ 77 F (25 C), H = 0, S + R ln P = 0

287

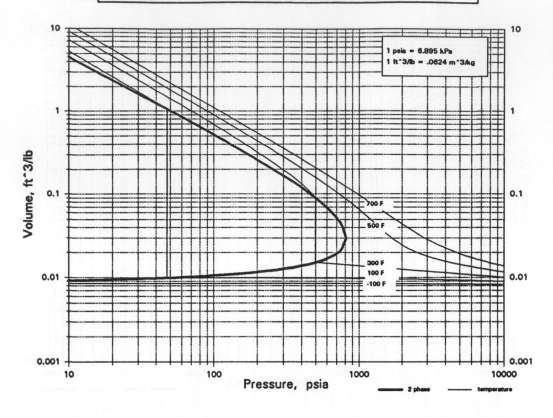

1 psia = 6.895 kPa
1 ft^3/lb = .0624 m^3/kg

Volume, ft^3/lb

Pressure, psia

━━ 2 phase   ── temperature

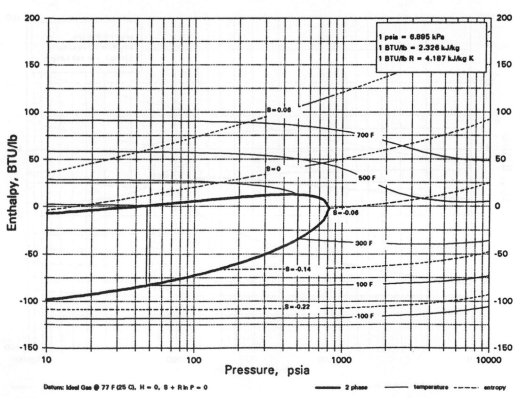

1 psia = 6.895 kPa
1 BTU/lb = 2.326 kJ/kg
1 BTU/lb R = 4.187 kJ/kg K

Enthalpy, BTU/lb

Pressure, psia

Datum: Ideal Gas @ 77 F (25 C), H = 0, S + R ln P = 0   ━━ 2 phase   ── temperature   ---- entropy

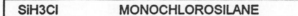

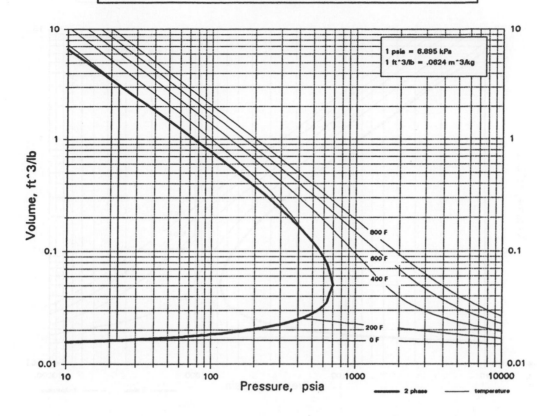

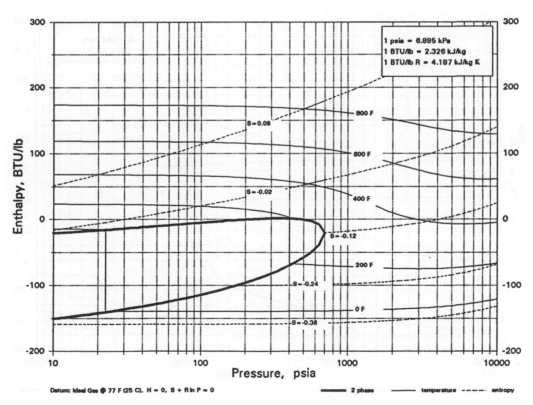

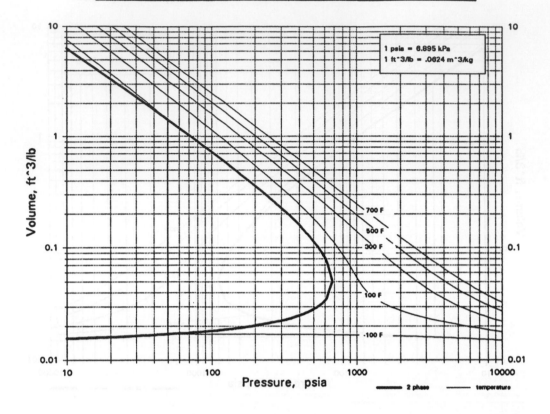

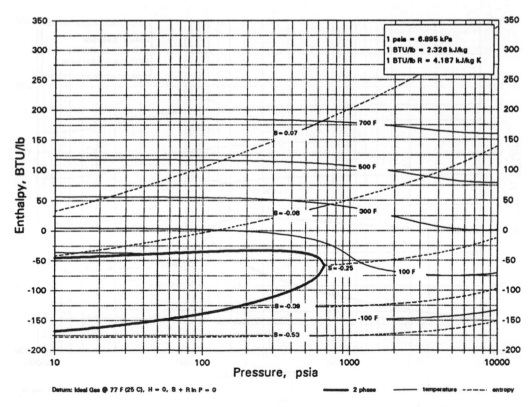

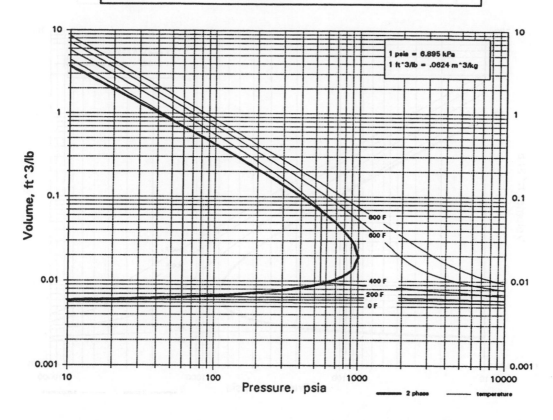

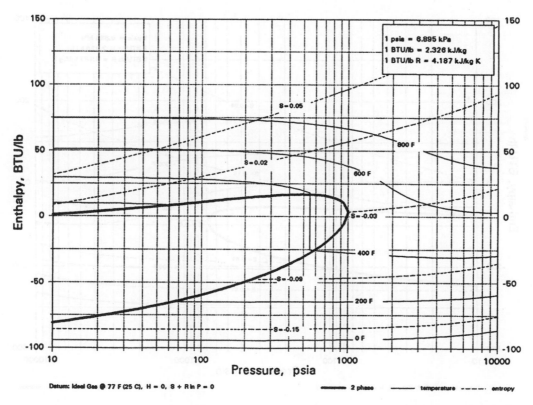

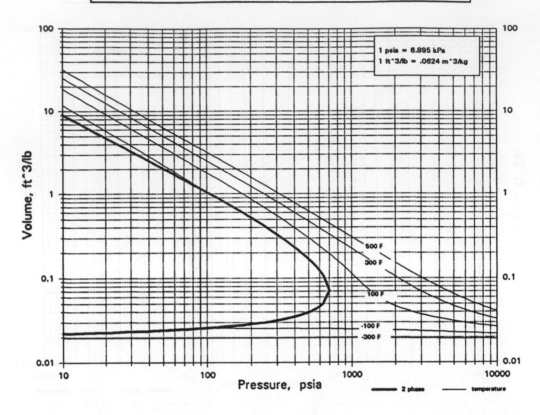

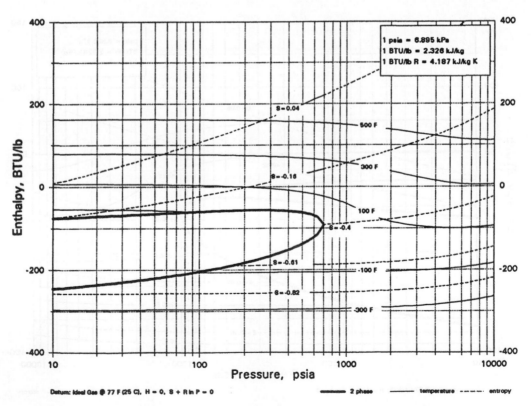

1. Molecular Weight, lb/mol.......... 60.084

2. Freezing Point, F........................ 2929.7

3. Boiling Point, F......................... 4046.1

4. Density @ 20 C, g/cm^3............ 2.66

5. Density @ 68 F, lb/ft^3.............. 166.06

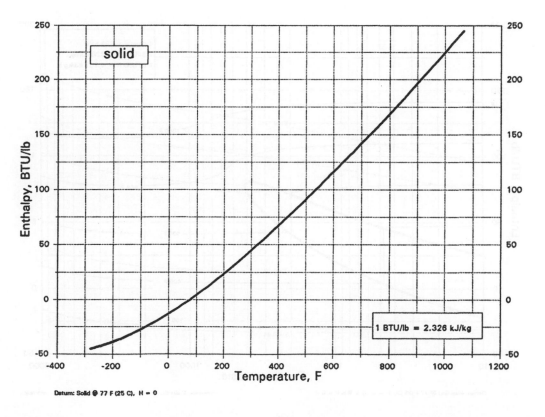

Datum: Solid @ 77 F (25 C), H = 0

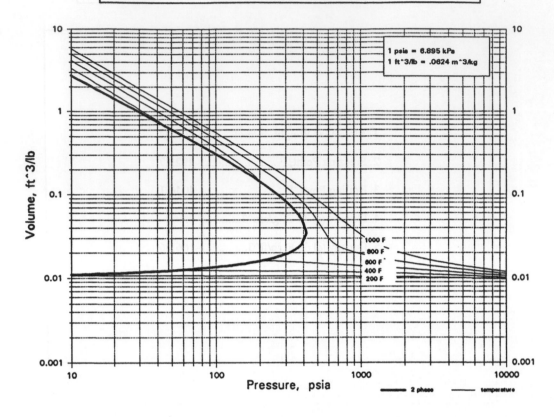

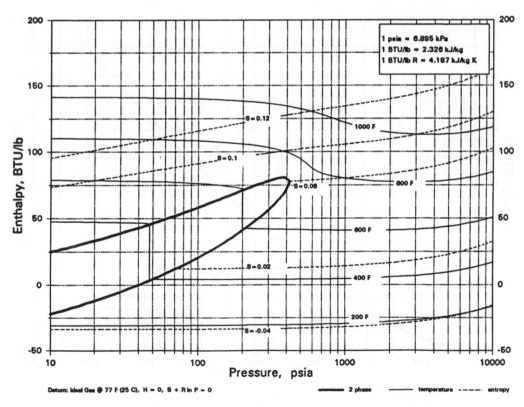

Datum: Ideal Gas @ 77 F (25 C), H = 0, S + R ln P = 0

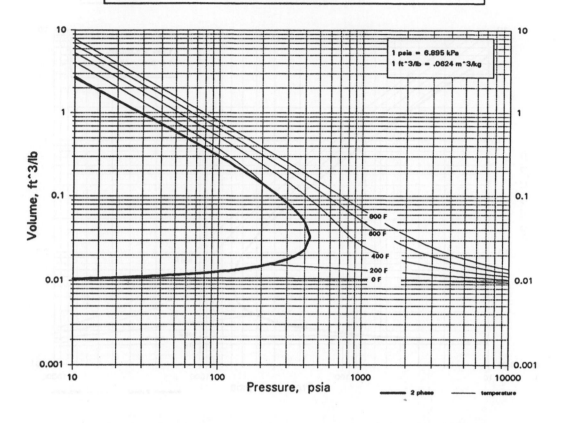

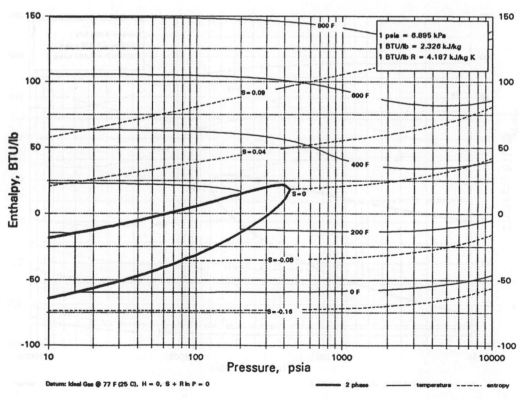

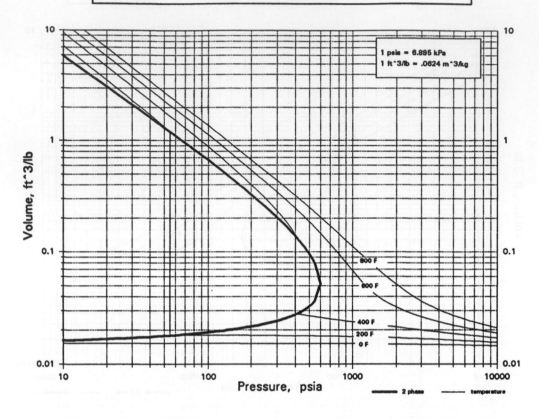

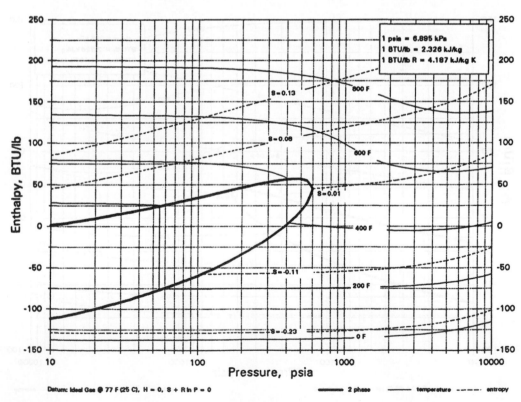

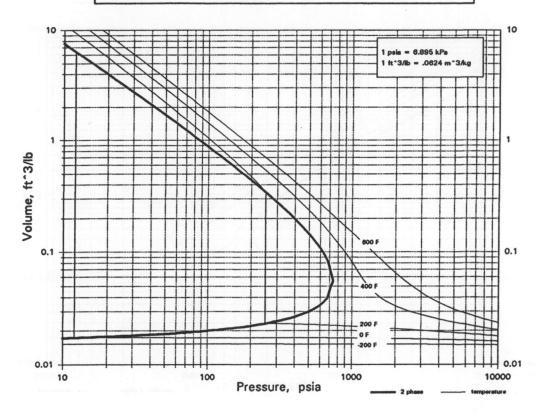

1 psia = 6.895 kPa
1 ft^3/lb = .0624 m^3/kg

600 F
400 F
200 F
0 F
-200 F

Volume, ft^3/lb

Pressure, psia

2 phase   temperature

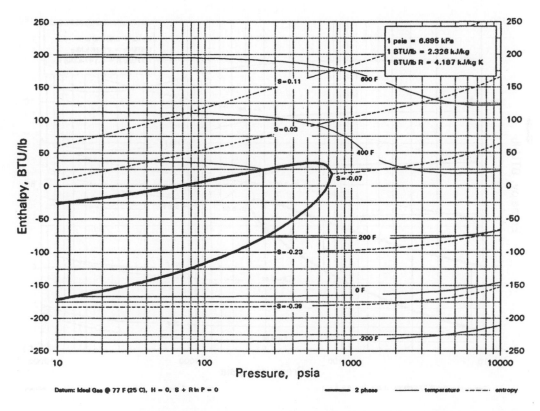

1 psia = 6.895 kPa
1 BTU/lb = 2.326 kJ/kg
1 BTU/lb R = 4.187 kJ/kg K

S=0.11
S=0.03
S=-0.07
S=-0.23
S=-0.39

600 F
400 F
200 F
0 F
-200 F

Enthalpy, BTU/lb

Pressure, psia

Datum: Ideal Gas @ 77 F (25 C), H = 0, S + R ln P = 0

2 phase   temperature   entropy

297

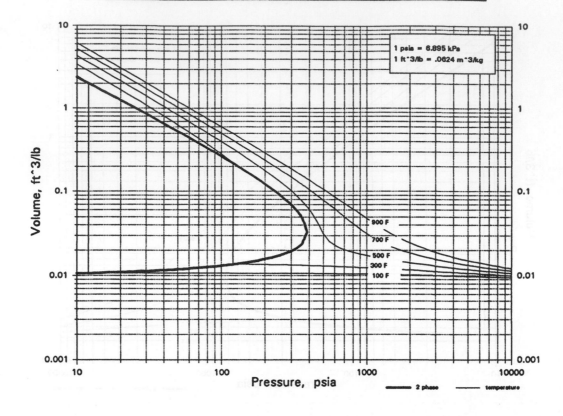

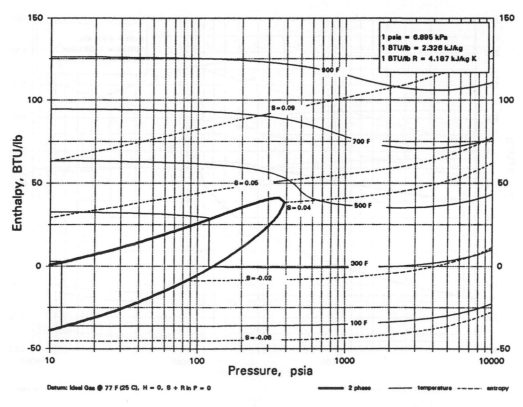

298

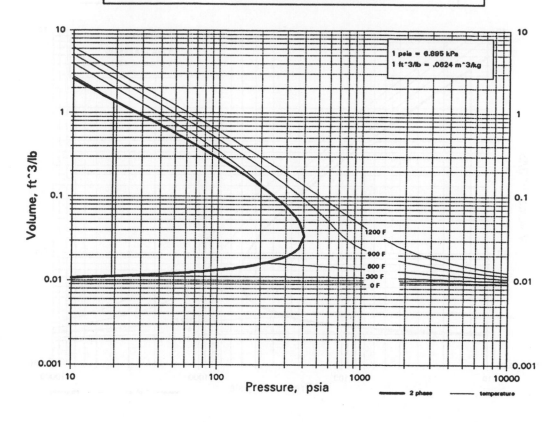

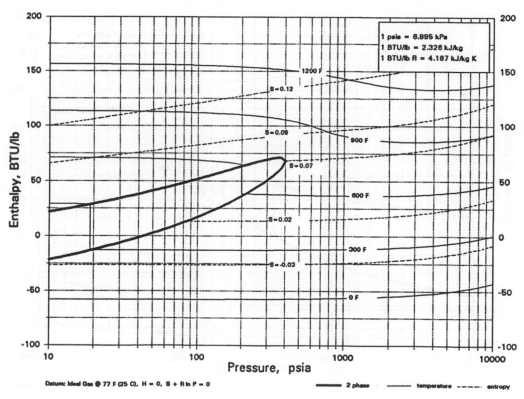

Datum: Ideal Gas @ 77 F (25 C), H = 0, S + R ln P = 0

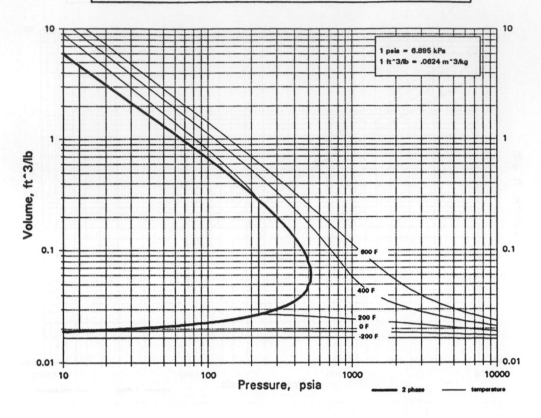

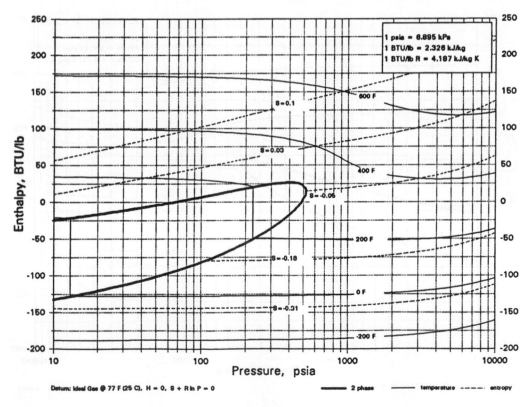

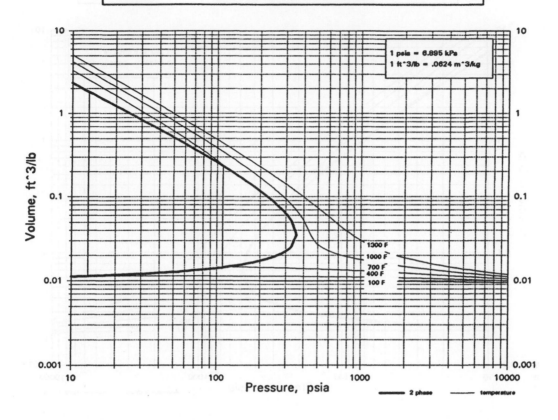

1 psia = 6.895 kPa
1 ft^3/lb = .0624 m^3/kg

1300 F
1000 F
700 F
400 F
100 F

━━━ 2 phase   ─── temperature

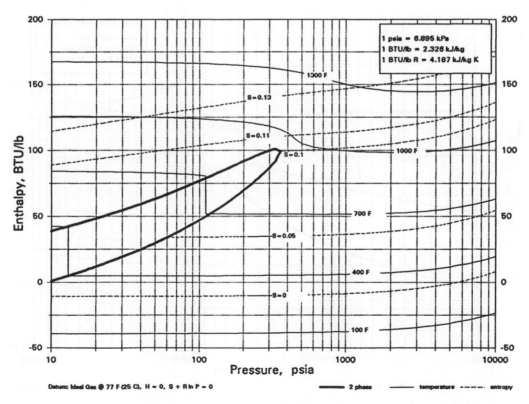

1 psia = 6.895 kPa
1 BTU/lb = 2.326 kJ/kg
1 BTU/lb R = 4.187 kJ/kg K

1300 F

S=0.13

S=0.11

S=0.1

1000 F

700 F

S=0.05

400 F

S=0

100 F

Datum: Ideal Gas @ 77 F (25 C), H = 0, S + R ln P = 0   ━━━ 2 phase   ─── temperature   ----- entropy

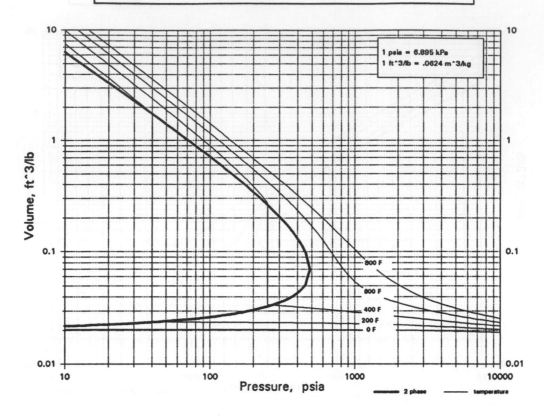

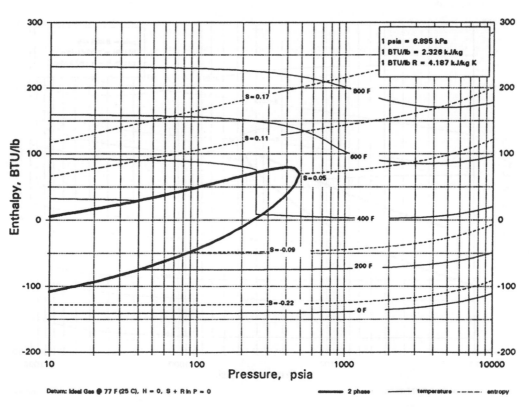

Datum: Ideal Gas @ 77 F (25 C), H = 0, S + R ln P = 0   2 phase   temperature   entropy

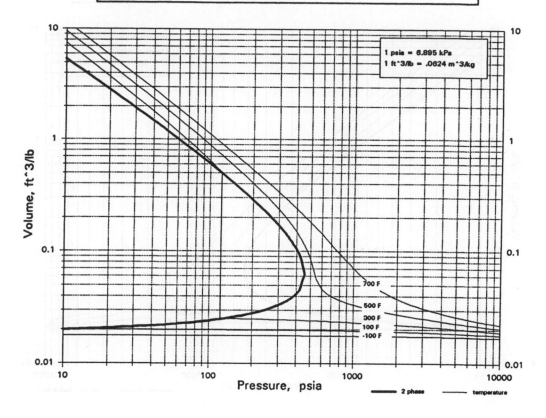

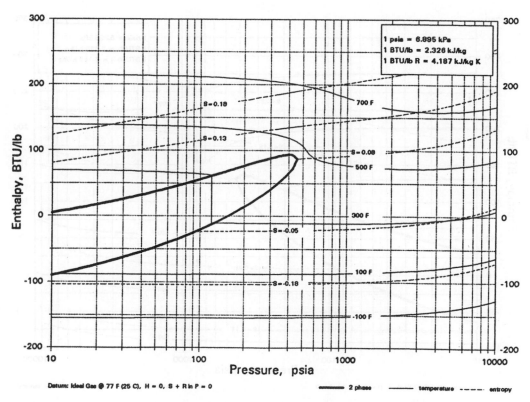

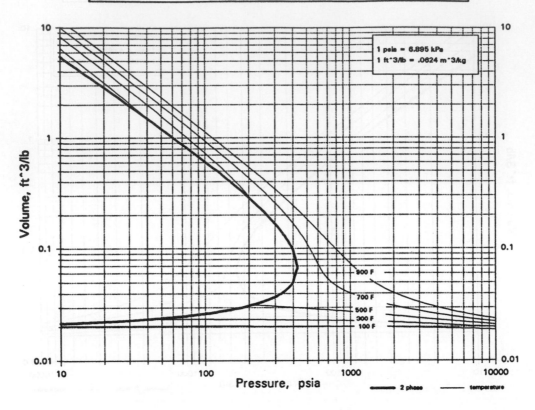

Volume, ft^3/lb

Pressure, psia

1 psia = 6.895 kPa
1 ft^3/lb = .0624 m^3/kg

900 F
700 F
500 F
300 F
100 F

2 phase   temperature

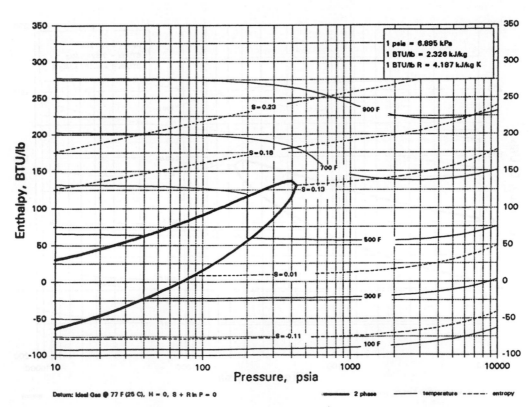

Enthalpy, BTU/lb

Pressure, psia

1 psia = 6.895 kPa
1 BTU/lb = 2.326 kJ/kg
1 BTU/lb R = 4.187 kJ/kg K

S = 0.23
S = 0.18
S = 0.13
S = 0.01
S = -0.11

900 F
700 F
500 F
300 F
100 F

Datum: Ideal Gas @ 77 F (25 C), H = 0, S + R ln P = 0

2 phase   temperature   - - - entropy

1. Molecular Weight, lb/mol.......... 150.36

2. Freezing Point, F........................ 1961.6

3. Boiling Point, F.......................... 2913.5

4. Density @ 20 C, g/cm^3............. 7.52

5. Density @ 68 F, lb/ft^3.............. 469.46

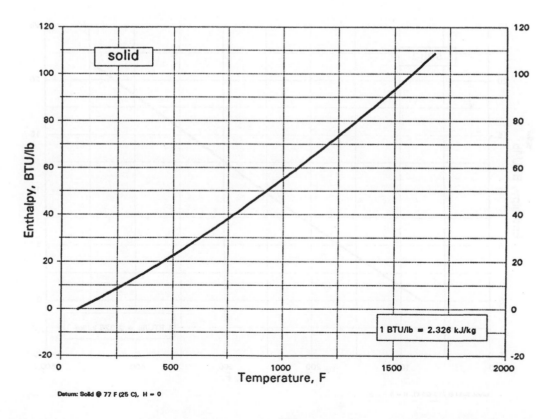

solid

1 BTU/lb = 2.326 kJ/kg

Datum: Solid @ 77 F (25 C), H = 0

1. Molecular Weight, lb/mol.......... 118.71

2. Freezing Point, F........................ 449.4

3. Boiling Point, F......................... 4931.3

4. Density @ 20 C, g/cm^3............ 7.28

5. Density @ 68 F, lb/ft^3.............. 454.47

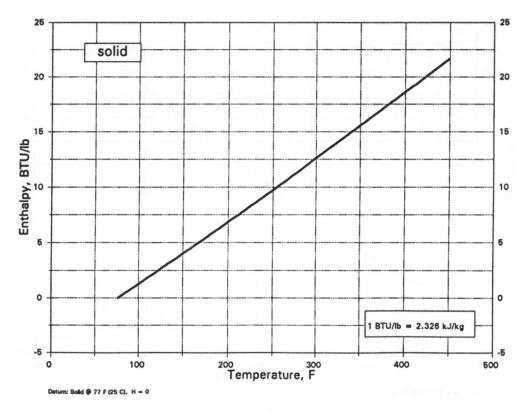

solid

1 BTU/lb = 2.326 kJ/kg

Datum: Solid @ 77 F (25 C), H = 0

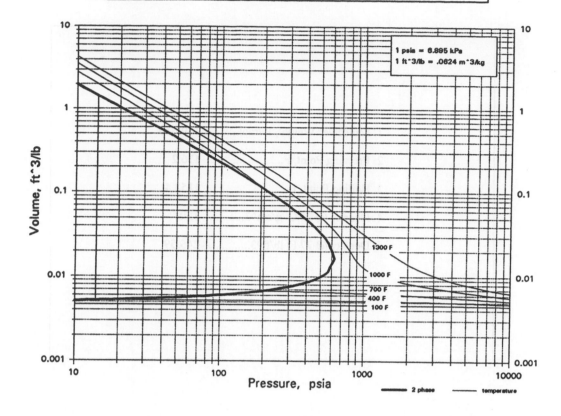

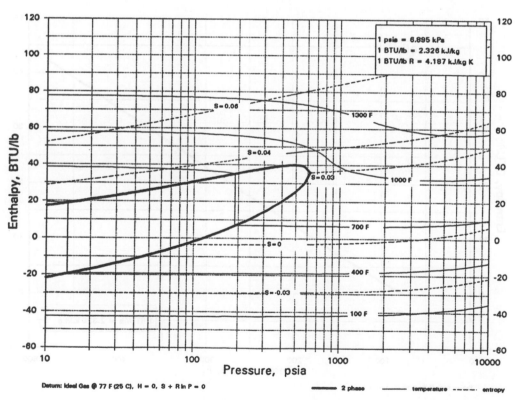

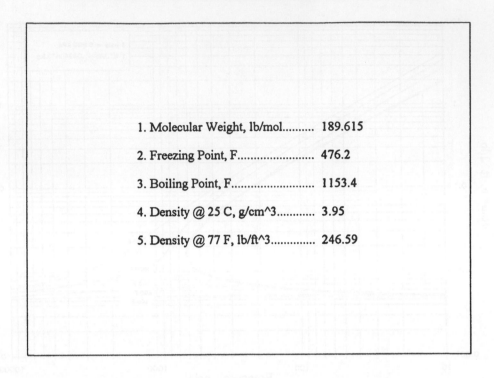

1. Molecular Weight, lb/mol.......... 189.615

2. Freezing Point, F........................ 476.2

3. Boiling Point, F........................ 1153.4

4. Density @ 25 C, g/cm^3............ 3.95

5. Density @ 77 F, lb/ft^3.............. 246.59

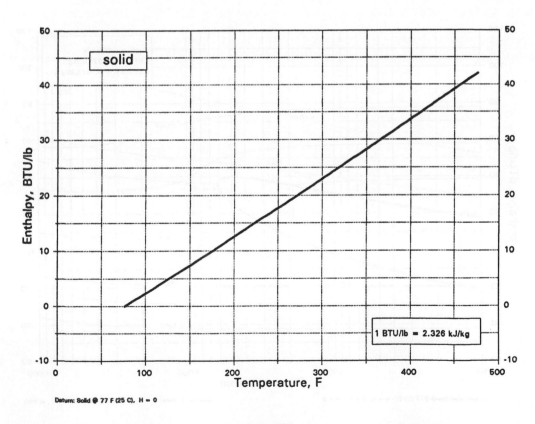

Datum: Solid @ 77 F (25 C). H = 0

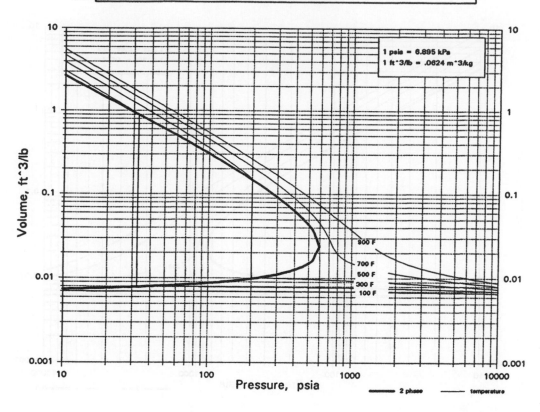

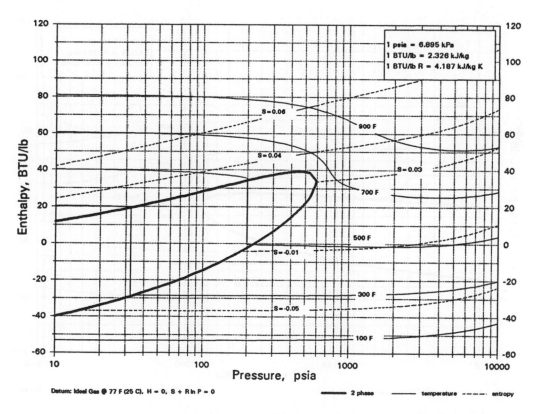

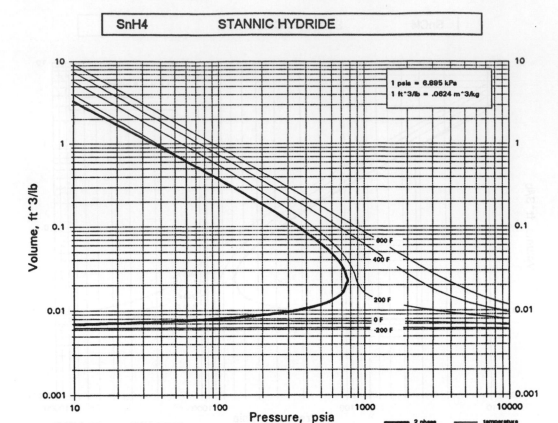

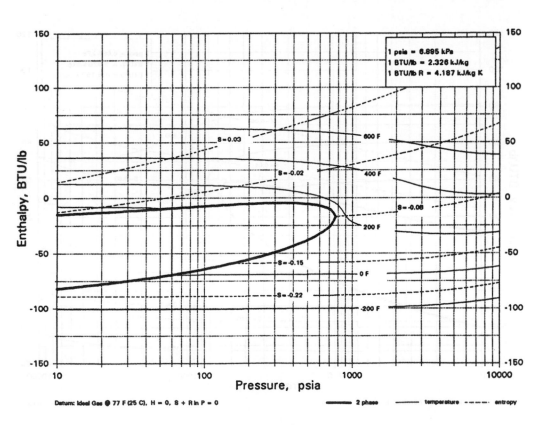

1. Molecular Weight, lb/mol.......... 626.328

2. Freezing Point, F....................... 292.1

3. Boiling Point, F........................ 658.4

4. Density @ 0 C, g/cm^3............ 4.47

5. Density @ 32 F, lb/ft^3.............. 279.05

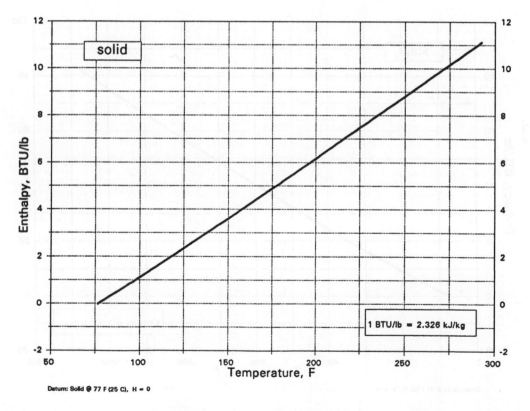

Datum: Solid @ 77 F (25 C), H = 0

1 BTU/lb = 2.326 kJ/kg

1. Molecular Weight, lb/mol.......... 87.62

2. Freezing Point, F........................ 1430.6

3. Boiling Point, F.......................... 2474.3

4. Density @ 20 C, g/cm^3............ 2.6

5. Density @ 68 F, lb/ft^3.............. 162.31

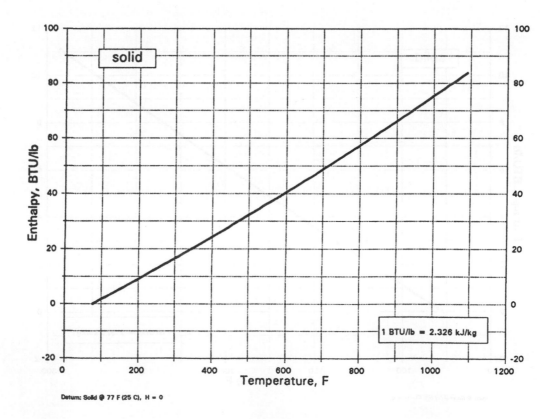

Datum: Solid @ 77 F (25 C), H = 0

312

1. Molecular Weight, lb/mol.......... 103.619

2. Freezing Point, F........................ 4406

3. Boiling Point, F......................... ---

4. Density @ 20 C, g/cm^3............. 4.7

5. Density @ 68 F, lb/ft^3.............. 293.41

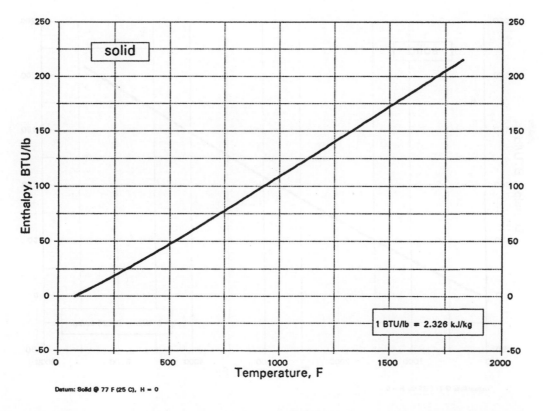

Datum: Solid @ 77 F (25 C), H = 0

313

1. Molecular Weight, lb/mol.......... 180.948

2. Freezing Point, F........................ 5462.6

3. Boiling Point, F.......................... 9557.3

4. Density @ 20 C, g/cm^3............ 16.6

5. Density @ 68 F, lb/ft^3.............. 1036.3

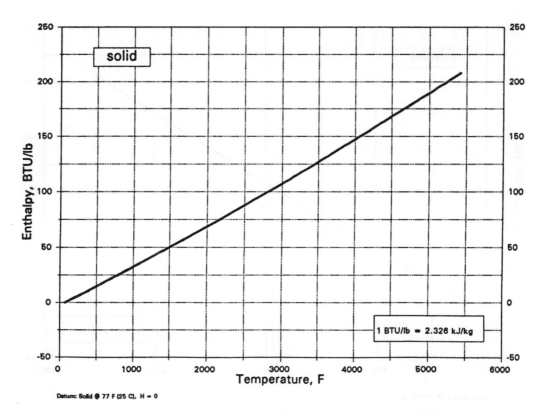

Datum: Solid @ 77 F (25 C), H = 0

1. Molecular Weight, lb/mol.......... 98

2. Freezing Point, F........................ 3914.6

3. Boiling Point, F........................ 8540.3

4. Density @ C, g/cm^3............. ---

5. Density @ F, lb/ft^3.............. ---

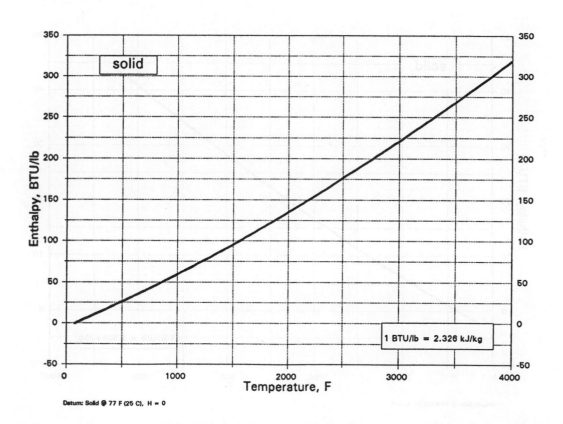

solid

1 BTU/lb = 2.326 kJ/kg

Datum: Solid @ 77 F (25 C), H = 0

315

1. Molecular Weight, lb/mol.......... 127.6

2. Freezing Point, F....................... 841.1

3. Boiling Point, F......................... 1853.3

4. Density @ 20 C, g/cm^3............ 6

5. Density @ 68 F, lb/ft^3.............. 374.57

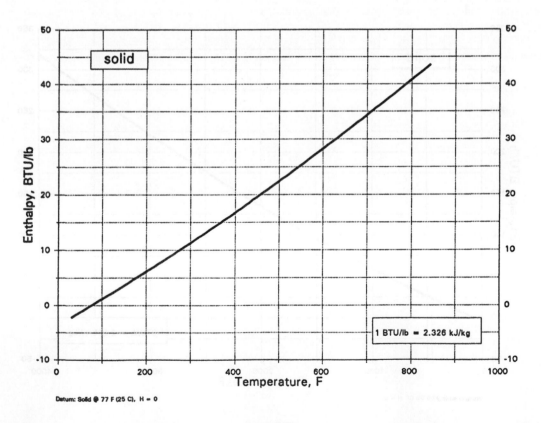

solid

1 BTU/lb = 2.326 kJ/kg

Datum: Solid @ 77 F (25 C), H = 0

1. Molecular Weight, lb/mol.......... 269.411

2. Freezing Point, F...................... 435.2

3. Boiling Point, F........................ 737.6

4. Density @ 18 C, g/cm^3............ 3.26

5. Density @ 64 F, lb/ft^3.............. 203.51

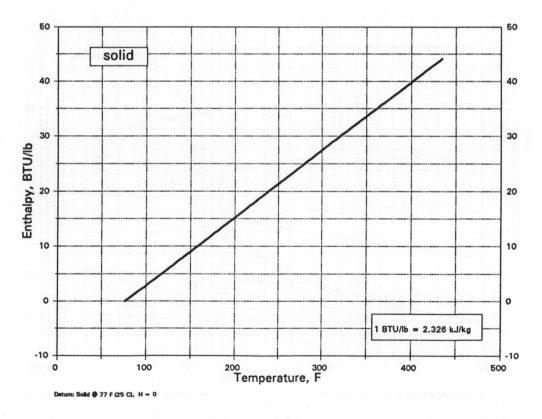

Datum: Solid @ 77 F (25 C), H = 0

317

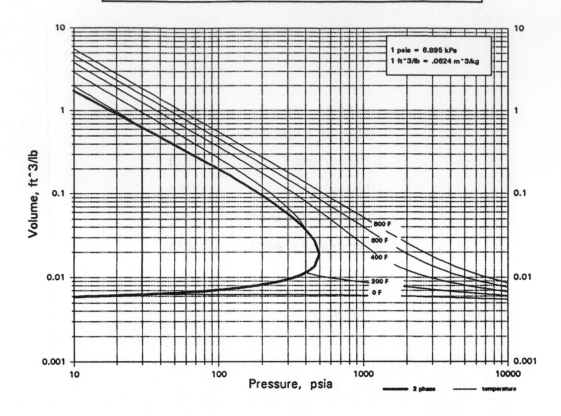

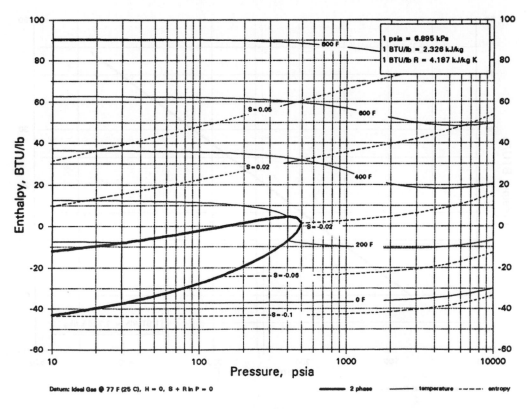

1. Molecular Weight, lb/mol.......... 47.88

2. Freezing Point, F........................ 3034.4

3. Boiling Point, F........................ 5735.9

4. Density @ 20 C, g/cm^3............ 4.5

5. Density @ 68 F, lb/ft^3.............. 280.93

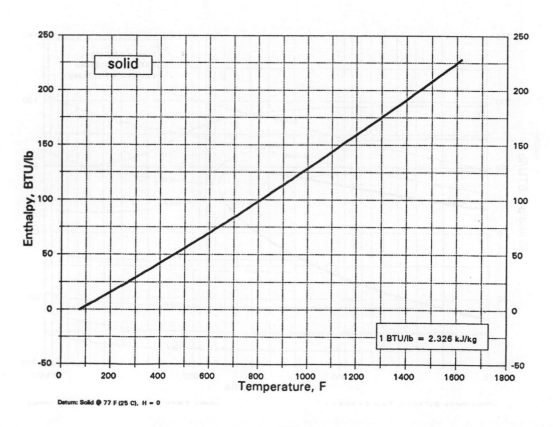

Datum: Solid @ 77 F (25 C), H = 0

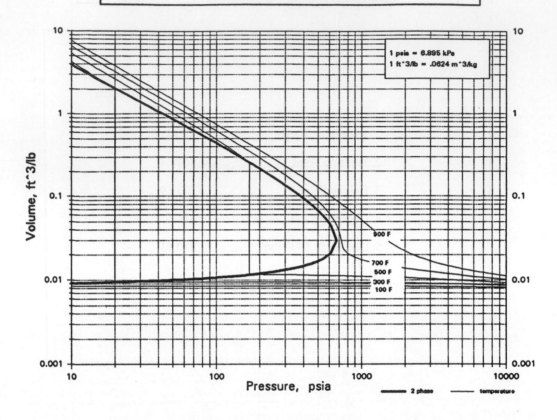

1 psia = 6.895 kPa
1 ft^3/lb = .0624 m^3/kg

900 F
700 F
500 F
300 F
100 F

2 phase   temperature

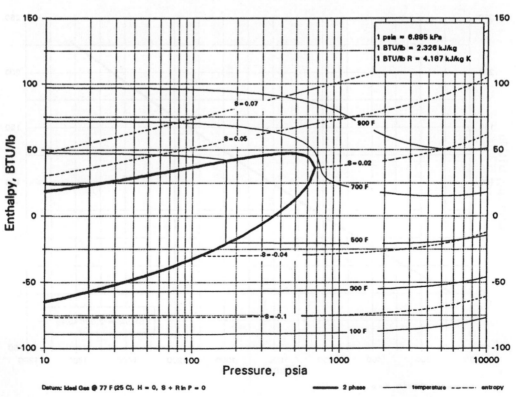

1 psia = 6.895 kPa
1 BTU/lb = 2.326 kJ/kg
1 BTU/lb R = 4.187 kJ/kg K

S = 0.07
S = 0.05
900 F
S = 0.02
700 F
S = -0.04
500 F
300 F
S = -0.1
100 F

Datum: Ideal Gas @ 77 F (25 C), H = 0, S + R ln P = 0   2 phase   temperature ----- entropy

1. Molecular Weight, lb/mol.......... 204.383

2. Freezing Point, F........................ 579.2

3. Boiling Point, F......................... 2681.3

4. Density @ 20 C, g/cm^3............ 11.85

5. Density @ 68 F, lb/ft^3.............. 739.77

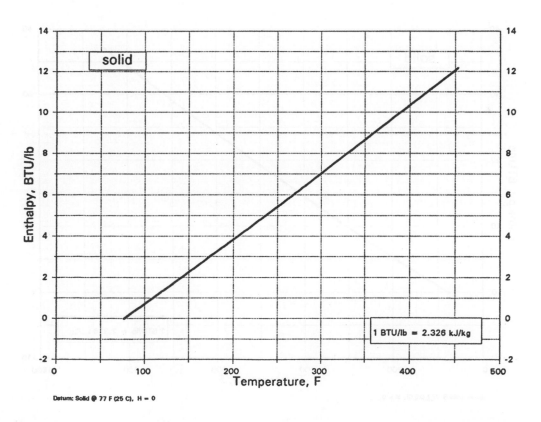

solid

1 BTU/lb = 2.326 kJ/kg

Enthalpy, BTU/lb

Temperature, F

Datum: Solid @ 77 F (25 C), H = 0

1. Molecular Weight, lb/mol.......... 284.287

2. Freezing Point, F........................ 860

3. Boiling Point, F.......................... 1506.2

4. Density @ 17 C, g/cm^3............ 7.56

5. Density @ 63 F, lb/ft^3.............. 471.95

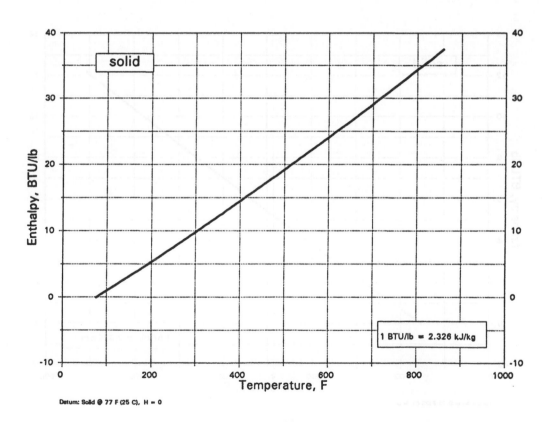

solid

1 BTU/lb = 2.326 kJ/kg

Enthalpy, BTU/lb

Temperature, F

Datum: Solid @ 77 F (25 C), H = 0

1. Molecular Weight, lb/mol.......... 331.288

2. Freezing Point, F........................ 824

3. Boiling Point, F.......................... 1513.4

4. Density @ 15 C, g/cm^3............ 7.1

5. Density @ 59 F, lb/ft^3.............. 443.24

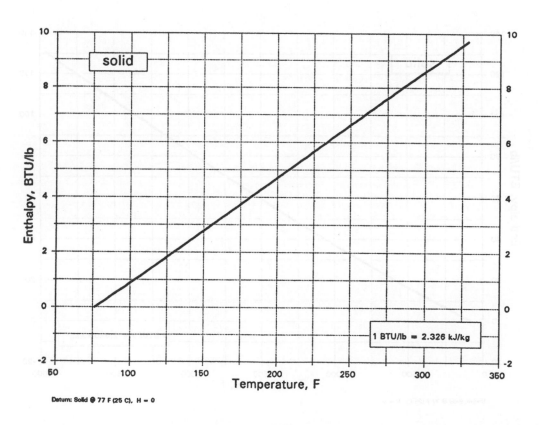

Datum: Solid @ 77 F (25 C), H = 0

1. Molecular Weight, lb/mol.......... 168.934

2. Freezing Point, F....................... 2813

3. Boiling Point, F......................... 3534.8

4. Density @ 20 C, g/cm^3............ 9.32

5. Density @ 68 F, lb/ft^3.............. 581.83

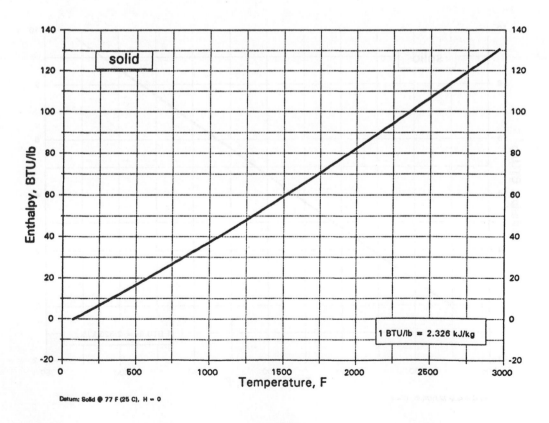

solid

1 BTU/lb = 2.326 kJ/kg

Datum: Solid @ 77 F (25 C), H = 0

1. Molecular Weight, lb/mol.......... 238.029

2. Freezing Point, F....................... 2075

3. Boiling Point, F......................... 6983.3

4. Density @ 25 C, g/cm^3............. 19.05

5. Density @ 77 F, lb/ft^3.............. 1189.25

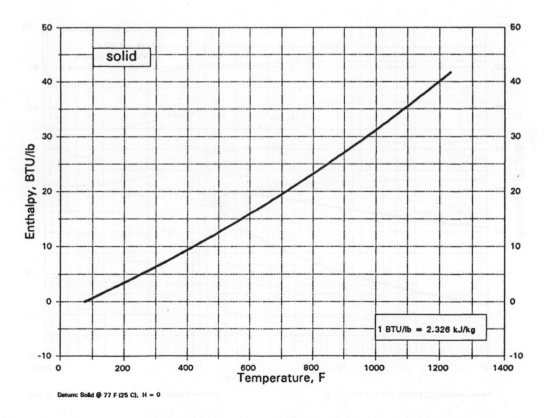

solid

1 BTU/lb = 2.326 kJ/kg

Datum: Solid @ 77 F (25 C), H = 0

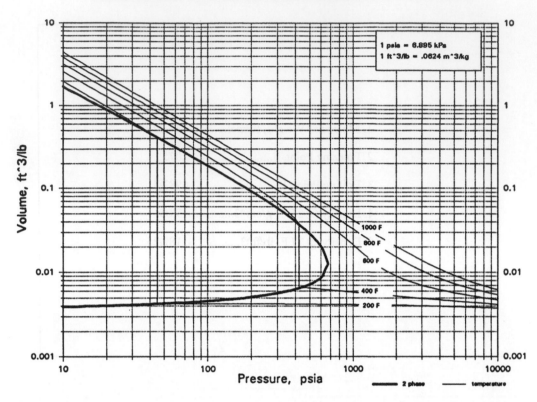

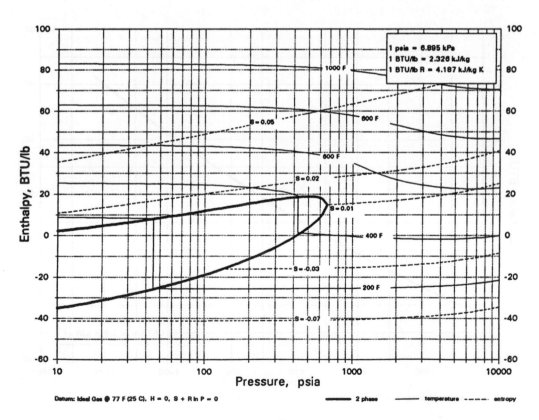

1. Molecular Weight, lb/mol.......... 50.942

2. Freezing Point, F........................ 3470

3. Boiling Point, F........................ 6137.3

4. Density @ 20 C, g/cm^3............ 5.96

5. Density @ 68 F, lb/ft^3.............. 372.07

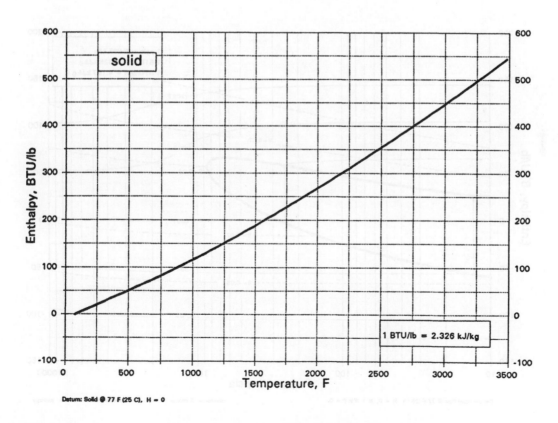

solid

1 BTU/lb = 2.326 kJ/kg

Datum: Solid @ 77 F (25 C), H = 0

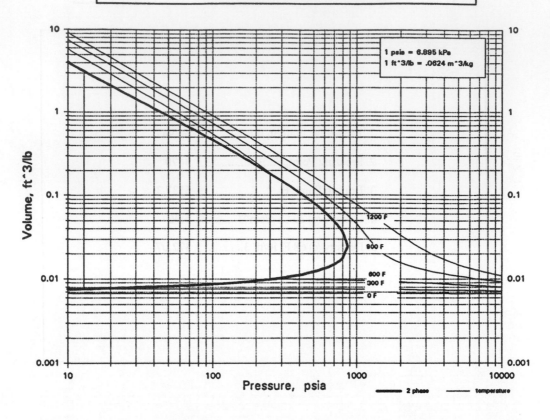

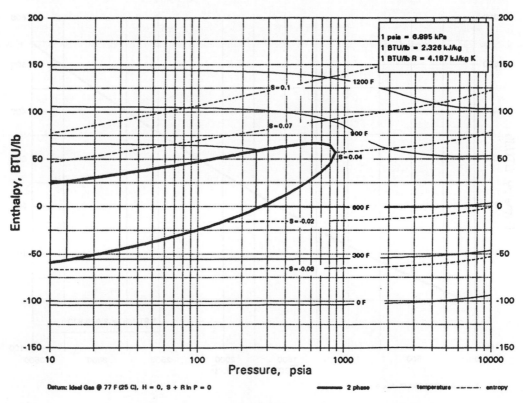

VOCl3   VANADIUM OXYTRICHLORIDE

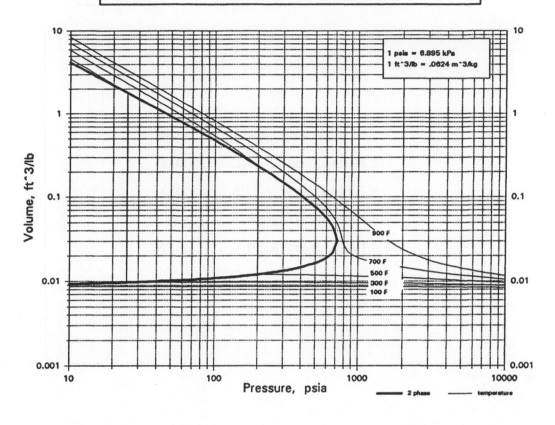

1 psia = 6.895 kPa
1 ft^3/lb = .0624 m^3/kg

900 F
700 F
500 F
300 F
100 F

Volume, ft^3/lb

Pressure, psia

2 phase   temperature

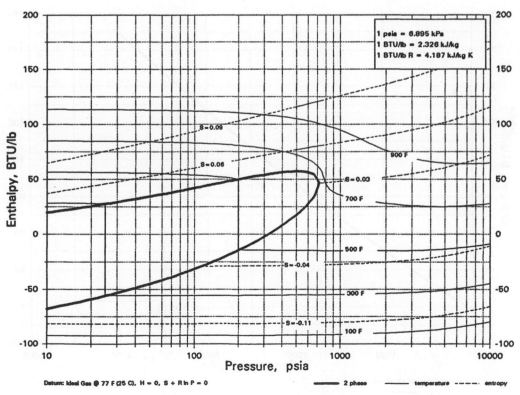

1 psia = 6.895 kPa
1 BTU/lb = 2.326 kJ/kg
1 BTU/lb R = 4.187 kJ/kg K

S = 0.09
S = 0.06
S = 0.03
S = -0.04
S = -0.11

900 F
700 F
500 F
300 F
100 F

Enthalpy, BTU/lb

Pressure, psia

Datum: Ideal Gas @ 77 F (25 C), H = 0, S + R ln P = 0

2 phase   temperature   entropy

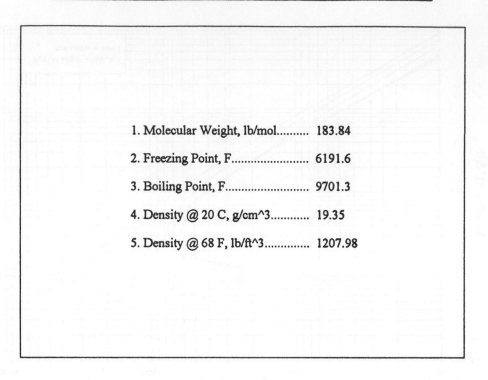

1. Molecular Weight, lb/mol.......... 183.84

2. Freezing Point, F...................... 6191.6

3. Boiling Point, F......................... 9701.3

4. Density @ 20 C, g/cm^3............ 19.35

5. Density @ 68 F, lb/ft^3.............. 1207.98

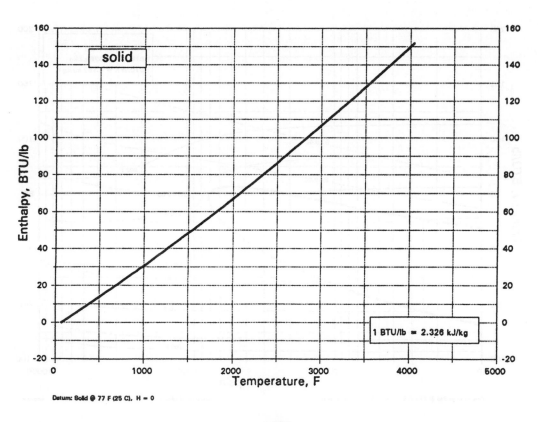

Datum: Solid @ 77 F (25 C), H = 0

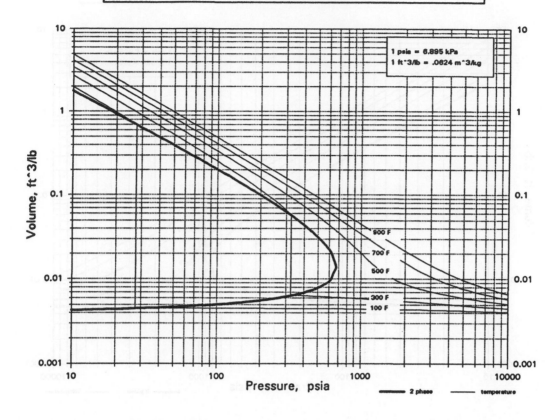

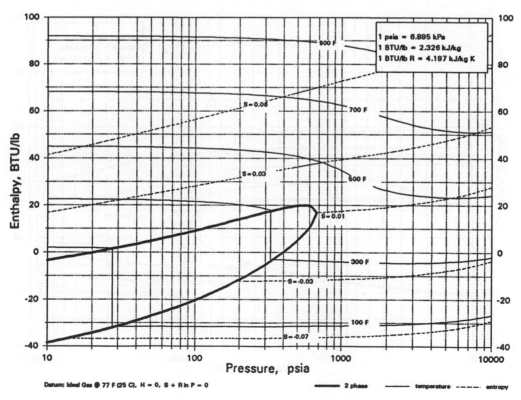

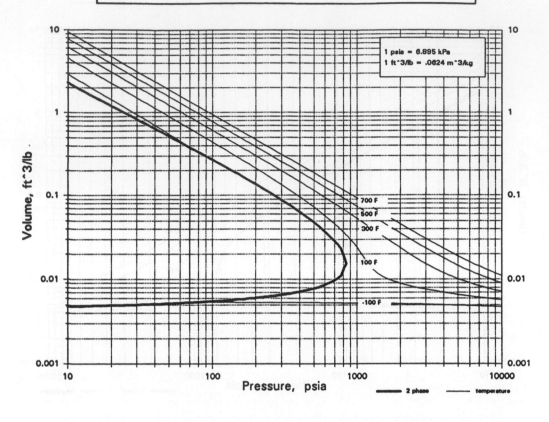

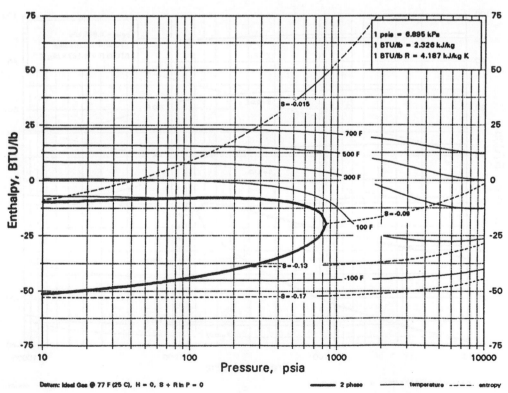

1. Molecular Weight, lb/mol.......... 173.04

2. Freezing Point, F....................... 1515.2

3. Boiling Point, F......................... 2528.3

4. Density @ 20 C, g/cm^3............. 6.97

5. Density @ 68 F, lb/ft^3.............. 435.12

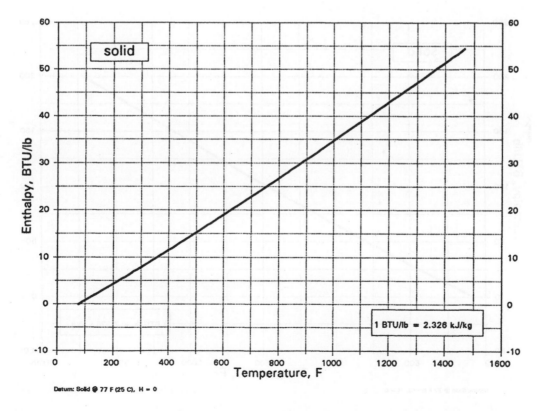

1 BTU/lb = 2.326 kJ/kg

Datum: Solid @ 77 F (25 C), H = 0

1. Molecular Weight, lb/mol.......... 88.906

2. Freezing Point, F...................... 2778.8

3. Boiling Point, F......................... 5039.3

4. Density @ 20 C, g/cm^3............ 4.47

5. Density @ 68 F, lb/ft^3.............. 279.05

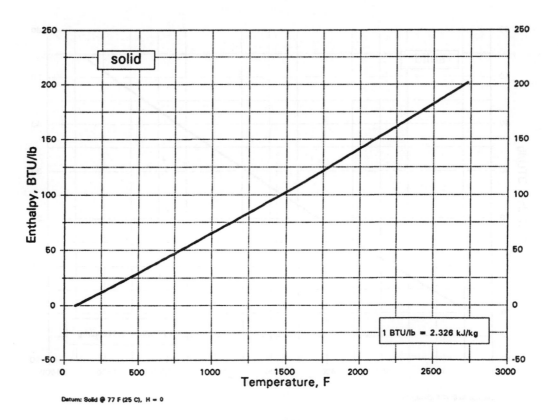

Datum: Solid @ 77 F (25 C), H = 0

1. Molecular Weight, lb/mol.......... 65.39

2. Freezing Point, F....................... 787.2

3. Boiling Point, F.......................... 1666.4

4. Density @ 20 C, g/cm^3............ 7.14

5. Density @ 68 F, lb/ft^3.............. 445.73

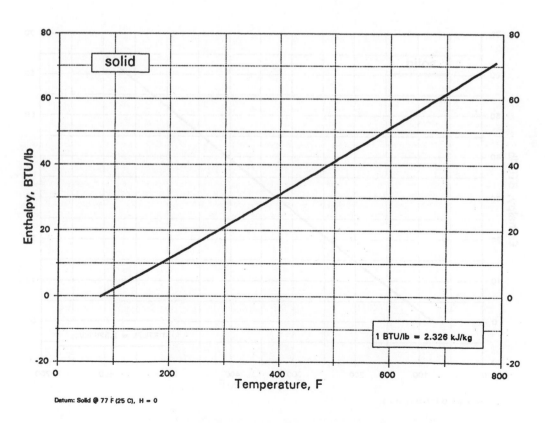

Datum: Solid @ 77 F (25 C), H = 0

1. Molecular Weight, lb/mol.......... 136.295

2. Freezing Point, F........................ 689

3. Boiling Point, F.......................... 1349.6

4. Density @ 25 C, g/cm^3............ 2.91

5. Density @ 77 F, lb/ft^3.............. 181.66

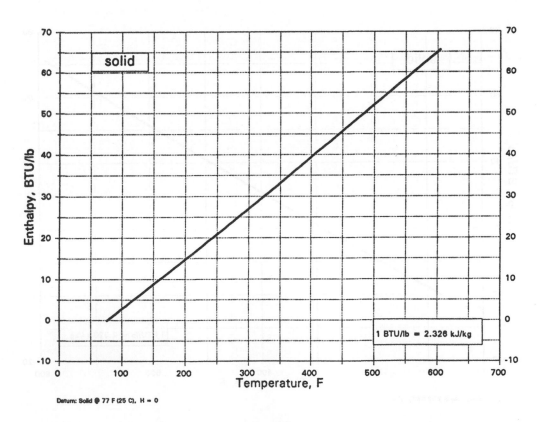

solid

1 BTU/lb = 2.326 kJ/kg

Datum: Solid @ 77 F (25 C), H = 0

1. Molecular Weight, lb/mol.......... 103.387

2. Freezing Point, F........................ 1601.6

3. Boiling Point, F.......................... 2726.6

4. Density @ 25 C, g/cm^3............ 4.95

5. Density @ 77 F, lb/ft^3.............. 309.02

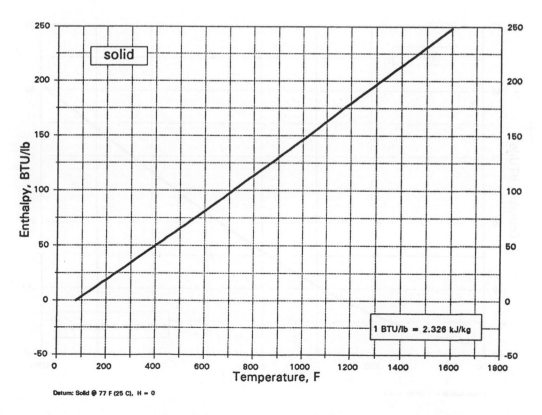

Datum: Solid @ 77 F (25 C), H = 0

1. Molecular Weight, lb/mol.......... 81.389

2. Freezing Point, F....................... 3587.1

3. Boiling Point, F.......................... ---

4. Density @ 20 C, g/cm^3............ 5.61

5. Density @ 68 F, lb/ft^3.............. 350.22

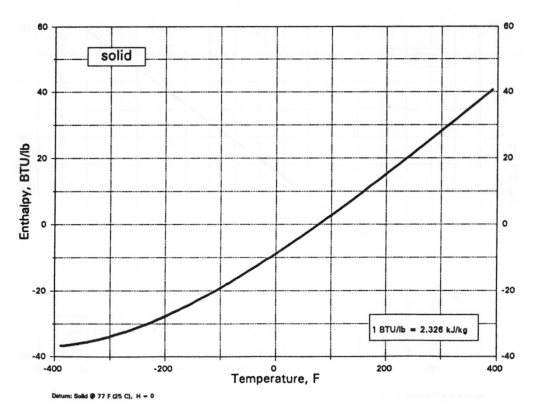

solid

1 BTU/lb = 2.326 kJ/kg

Datum: Solid @ 77 F (25 C), H = 0

1. Molecular Weight, lb/mol.......... 161.454

2. Freezing Point, F........................ 1255.7

3. Boiling Point, F........................ ---

4. Density @ 25 C, g/cm^3............. 3.54

5. Density @ 77 F, lb/ft^3.............. 220.99

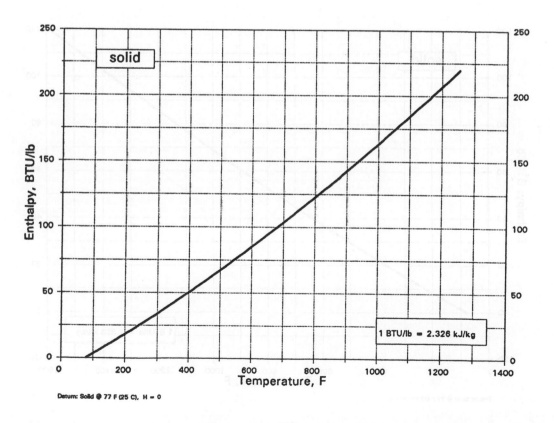

Datum: Solid @ 77 F (25 C), H = 0

1. Molecular Weight, lb/mol.......... 91.224

2. Freezing Point, F...................... 3371

3. Boiling Point, F......................... 7816.7

4. Density @ 20 C, g/cm^3............ 6.49

5. Density @ 68 F, lb/ft^3.............. 405.16

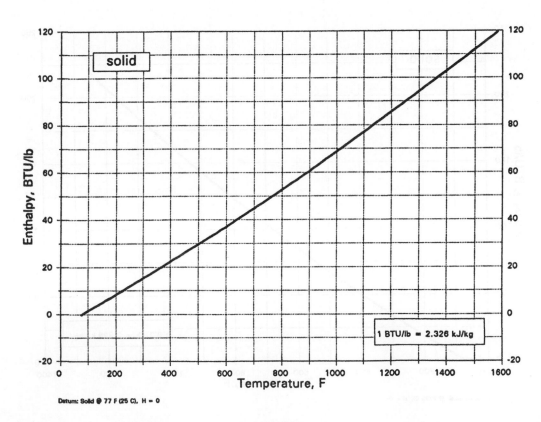

solid

Enthalpy, BTU/lb

Temperature, F

1 BTU/lb = 2.326 kJ/kg

Datum: Solid @ 77 F (25 C), H = 0

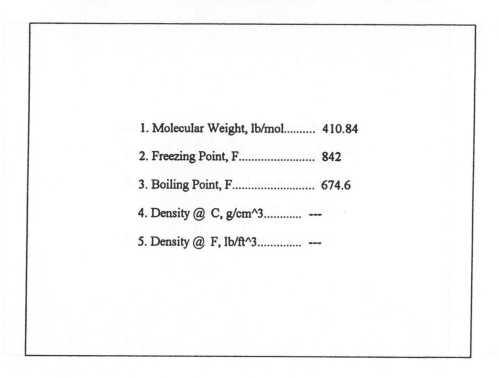

1. Molecular Weight, lb/mol.......... 410.84

2. Freezing Point, F........................ 842

3. Boiling Point, F.......................... 674.6

4. Density @ C, g/cm^3............. ---

5. Density @ F, lb/ft^3.............. ---

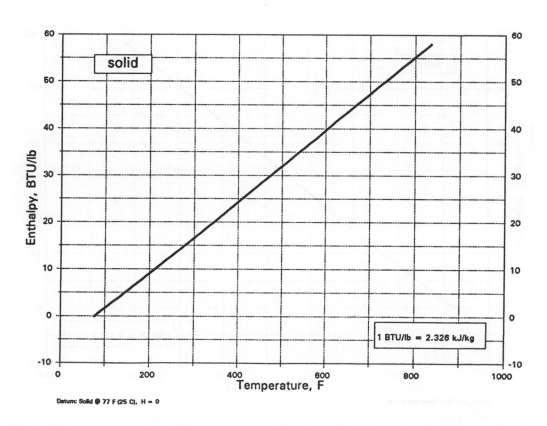

Datum: Solid @ 77 F (25 C), H = 0

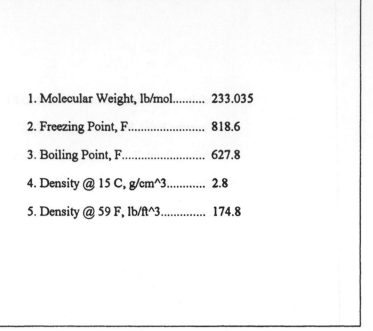

1. Molecular Weight, lb/mol........... 233.035

2. Freezing Point, F........................ 818.6

3. Boiling Point, F........................... 627.8

4. Density @ 15 C, g/cm^3............ 2.8

5. Density @ 59 F, lb/ft^3.............. 174.8

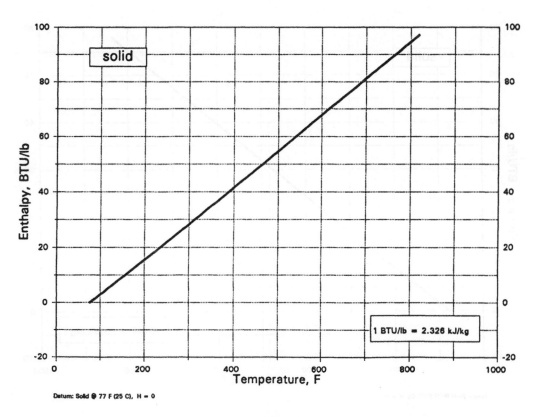

Datum: Solid @ 77 F (25 C), H = 0

1. Molecular Weight, lb/mol.......... 598.842

2. Freezing Point, F...................... 930.2

3. Boiling Point, F.......................... 807.8

4. Density @ C, g/cm^3............ ---

5. Density @ F, lb/ft^3.............. ---

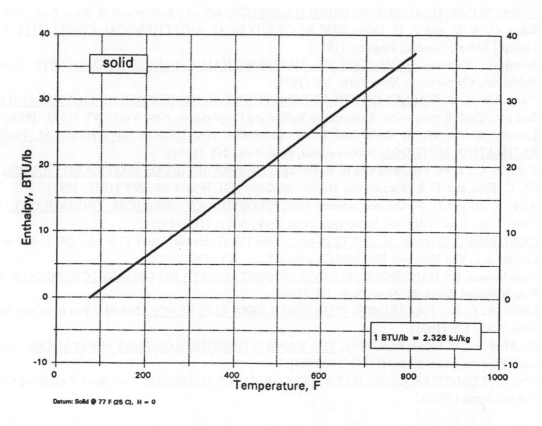

solid

1 BTU/lb = 2.326 kJ/kg

Enthalpy, BTU/lb

Temperature, F

Datum: Solid @ 77 F (25 C), H = 0

REFERENCES

1. Peng, D. Y. and D. B. Robinson, Ind. Eng. Chem. Fundam., $\underline{15}$ (No. 1), 59 (1976).
2. Stryjek, R. and J. H. Vera, Can. J. Chem. Eng., $\underline{64}$, 323 (1986).
3. Stryjek, R. and J. H. Vera, Can. J. Chem. Eng., $\underline{64}$, 334 (1986).
4. SELECTED VALUES OF PROPERTIES OF HYDROCARBONS AND RELATED COMPOUNDS, Thermodynamics Research Center, TAMU, College Station, TX (1977, 1984).
5. SELECTED VALUES OF PROPERTIES OF CHEMICAL COMPOUNDS, Thermodynamics Research Center, TAMU, College Station, TX (1977, 1987).
6. Ambrose, D., VAPOUR-LIQUID CRITICAL PROPERTIES, National Physical Laboratory, Teddington, England, NPL Report Chem 107 (Feb., 1980).
7. Nesmeyanov, A. N., VAPOR PRESSURE OF THE CHEMICAL ELEMENTS, Elsevier, New York, NY (1963).
8. Daubert, T. E. and R. P. Danner, DATA COMPILATION OF PROPERTIES OF PURE COMPOUNDS, Parts 1, 2, 3 and 4, Supplements 1 and 2, DIPPR Project, AIChE, New York, NY (1985-1992).
9. Simmrock, K. H., R. Janowsky and A. Ohnsorge, CRITICAL DATA OF PURE SUBSTANCES, Vol. II, Parts 1 and 2, Dechema Chemistry Data Series, 6000 Frankfurt/Main, Germany (1986).
10. INTERNATIONAL CRITICAL TABLES, McGraw-Hill, New York, NY (1926).
11. Braker, W. and A. L. Mossman, MATHESON GAS DATA BOOK, 6th ed., Matheson Gas Products, Secaucaus, NJ (1980).
12. CRC HANDBOOK OF CHEMISTRY AND PHYSICS, 66th - 75th eds., CRC Press, Inc., Boca Raton, FL (1985-1994).
13. LANGE'S HANDBOOK OF CHEMISTRY, 13th and 14th eds., McGraw-Hill, New York, NY (1985, 1992).
14. PERRY'S CHEMICAL ENGINEERING HANDBOOK, 6th ed., McGraw-Hill, New York, NY (1984).
15. Kaye, G. W. C. and T. H. Laby, TABLES OF PHYSICAL AND CHEMICAL CONSTANTS, Longman Group Limited, London, England (1973).
16. Raznjevic, Kuzman, HANDBOOK OF THERMODYNAMIC TABLES AND CHARTS, Hemisphere Publishing Corporation, New York, NY (1976).
17. Vargaftik, N. B., TABLES ON THE THERMOPHYSICAL PROPERTIES OF LIQUIDS AND GASES, 2nd ed., English translation, Hemisphere Publishing Corporation, New York, NY (1975, 1983).
18. Lyman, W. J., W. F. Reehl and D. H. Rosenblatt, HANDBOOK OF CHEMICAL PROPERTY ESTIMATION METHODS, McGraw-Hill, New York, NY (1982).
19. Reid, R. C., J. M. Prausnitz and B. E. Poling, THE PROPERTIES OF GASES AND LIQUIDS, 3rd ed. (R. C. Reid and T. K. Sherwood), 4th ed., McGraw-Hill, New York, NY (1977, 1987).
20. Kirk, R. E. and D. F. Othmer, editors, ENCYCLOPEDIA OF CHEMICAL TECHNOLOGY, 3rd ed., Vols. 1-24, John Wiley and Sons, Inc., New York, NY (1978-1984).
21. CONDENSED CHEMICAL DICTIONARY, 10th (G. G. Hawley) and 11th eds. (N. I. Sax and R. J. Lewis, Jr.), Van Nostrand Reinhold Co., New York, NY (1981,1987).
22. Verschueren, K., HANDBOOK OF ENVIRONMENTAL DATA ON ORGANIC CHEMICALS, 2nd ed., Van Nostrand Reinhold, New York, NY (1983).
23. Lewis, R. J., Sr., HAZARDOUS CHEMICALS DESK REFERENCE, 2nd ed., Van Nostrand Reinhold, New York, NY (1991).
24. Boublick, T., V. Fried and E. Hala, THE VAPOUR PRESSURES OF PURE SUBSTANCES, 1st and 2nd eds., Elsevier, New York, NY (1975, 1984).
25. Ohe, S., COMPUTER AIDED DATA BOOK OF VAPOR PRESSURE, Data Book Publishing Company, Tokyo, Japan (1976).

26. Hultgren, R., P. D. Desai, D. T. Hawkins, M. Gleiser, K. K. Kelley and D. D. Wagman, SELECTED VALUES OF THE THERMODYNAMIC PROPERTIES OF THE ELEMENTS, American Society for Metals, Metals Park, OH (1973).

27. Simmrock, K. H., R. Janowsky and A. Ohnsorge, CRITICAL DATA OF PURE SUBSTANCES, Vol. II, Parts 1 and 2, Dechema Chemistry Data Series, 6000 Frankfurt/Main, Germany (1986).

28. Stull, D. R. and H. Prophet, JANAF THERMOCHEMICAL TABLES, 2nd edition, NSRDS-NBS 37, US Government Printing Office, Washington, DC (June, 1971).

29. Wagman, D. D. and others, NBS TABLES OF CHEMICAL THERMODYNAMIC PROPERTIES, J. Phys. Chem. Ref. Data, 11, Supplement No. 2 (1982).

30. Chase, M. W. and others, JANAF THERMOCHEMICAL TABLES, 3rd edition, Parts 1 (Al-Co) and 2 (Cr-Zr), J. Phys. Chem. Ref. Data, 14, Supplement No. 1 (1985).

31. Kelley, K. K., CONTRIBUTIONS TO THE DATA ON THEORETICAL METALLURGY, Bureau of Mines Bulletin 584, US Government Printing Office, Washington, DC (1960).

32. Wicks, C. E. and F. E. Block, THERMODYNAMIC PROPERTIES OF 65 ELEMENTS - THEIR OXIDES, HALIDES, CARBIDES, AND NITRIDES, Bureau of Mines Bulletin 605, US Government Printing Office, Washington, DC (1963).

33. Barin, I. and O. Knacke, THERMOCHEMICAL PROPERTIES OF INORGANIC SUBSTANCES, Springer-Verlag, New York, NY (1973).

34. Yaws, C. L. and others, Solid State Technology, 16, No. 1, 39 (1973).

35. Yaws, C. L. and others, Solid State Technology, 17, No. 1, 47 (1974).

36. Yaws, C. L. and others, Solid State Technology, 17, No. 11, 31 (1974).

37. Yaws, C. L. and others, Solid State Technology, 18, No. 1, 35 (1975).

38. Yaws, C. L. and others, Solid State Technology, 21, No. 1, 43 (1978).

39. Yaws, C. L. and others, Solid State Technology, 24, No. 1, 87 (1981).

40. Yaws, C. L. and others, J. Ch. I. Ch. E., 12, 33 (1981).

41. Yaws, C. L. and others, J. Ch. I. Ch. E., 14, 205 (1983).

42. Yaws, C. L. and others, Ind. Eng. Chem. Process Des. Dev., 23, 48 (1984).

43. Yaws, C. L., PHYSICAL PROPERTIES, McGraw-Hill, New York, NY (1977).

44. Yaws, C. L., THERMODYNAMIC AND PHYSICAL PROPERTY DATA, Gulf Publishing Co., Houston, TX (1992).

45. Yaws, C. L. and R. W. Gallant, PHYSICAL PROPERTIES OF HYDROCARBONS, Vols. 1 (2nd ed.), 2 (3rd ed.), 3, and 4, Gulf Publishing Co., Houston, TX (1992, 1993, 1993, 1995).

46. Yaws, C. L., HANDBOOK OF VAPOR PRESSURE, Vol. 4 - Inorganic Compounds and Elements, Gulf Publishing Co., Houston, TX (1995).

Appendix A

Equations for Thermodynamic Properties

Enthalpy

$$H = H_{ref} + \int_{T_{ref}}^{T} C_P \, dT - \Delta H^{resid} \tag{1}$$

Entropy

$$S = S_{ref} + \int_{T_{ref}}^{T} \frac{C_P}{T} \, dT - R \ln\left(\frac{P}{P_{ref}}\right) - \Delta S^{resid} \tag{2}$$

Internal Energy

$$U = H - P V \tag{3}$$

Helmholtz Energy

$$A = U - T S \tag{4}$$

Gibbs Energy

$$G = H - T S \tag{5}$$

Parameters

$$C_P = heat\ capacity\ of\ ideal\ gas \tag{6}$$

$$H_{ref}, S_{ref} = reference\ state\ for\ ideal\ gas \tag{7}$$

$$T_{ref}, P_{ref} = reference\ temperature, reference\ pressure \tag{8}$$

$$\Delta H^{resid}, \Delta S^{resid} = residual\ enthalpy, residual\ entropy \tag{9}$$

Appendix B

Peng-Robinson Equation of State for Thermodynamic Properties

Equation of State

$$P = \frac{RT}{V - b} - \frac{a}{V(V + b) + b(V - b)} \tag{1}$$

Volume

$$V^3 + (b - \frac{RT}{P})V^2 + (\frac{a}{P} - 3b^2 - \frac{RT}{P}2b)V + (b^3 + \frac{RT}{P}b^2 - \frac{ab}{P}) = 0 \tag{2}$$

Compressibility Factor

$$Z^3 - (1 - B)Z^2 + (A - 3B^2 - 2B)Z - (AB - B^2 - B^3) = 0 \tag{3}$$

Fugacity Coefficient

$$\ln \phi = Z - 1 - \ln(Z - B) - \frac{A}{2\sqrt{2}B}\ln\left(\frac{Z + 2.414B}{Z - 0.414B}\right) \tag{4}$$

Residual Enthalpy

$$\frac{\Delta H^{resid}}{RT} = 1 - Z + \frac{A}{2\sqrt{2}B}(1 + \frac{D}{a})\ln\left(\frac{Z + 2.414B}{Z - 0.414B}\right) \tag{5}$$

Residual Entropy

$$\frac{\Delta S^{resid}}{R} = -\ln(Z - B) + \frac{AD}{2\sqrt{2}Ba}\ln\left(\frac{Z + 2.414B}{Z - 0.414B}\right) \tag{6}$$

Parameters

$$a = a_c\alpha \tag{7}$$

$$a_c = 0.45724R^2T_c^2/P_c \tag{8}$$

$$b = 0.07780RT_c/P_c \tag{9}$$

$$\alpha = [1 + m(1 - T_r^{1/2})]^2 \tag{10}$$

$$m = 0.37464 + 1.54226\Omega - 0.26992\Omega^2 \quad (original\ PR) \tag{11}$$

$$m = see\ Stryjek,\ Vera\ (modified\ PR) \tag{12}$$

$$A = aP/R^2T^2 = 0.45724\alpha P_r/T_r^2 = 0.45724\frac{(P/P_c)}{(T/T_c)^2}\alpha \tag{13}$$

$$B = bP/RT = 0.07780P_r/T_r = 0.07780\frac{(P/P_c)}{(T/T_c)} \tag{14}$$

$$D = -T\frac{da}{dT} = ma\sqrt{T_r/\alpha} \tag{15}$$

Appendix C

Examples for Thermodynamic Diagrams

Example 1 - Vessel Pressure

A vessel containing gaseous chlorine (Cl2) at 395 psia and 200 F is exposed to a fire in the process area. The temperature in the vessel is 800 F when the fire is extinguished. Estimate the final pressure in the vessel.

Since the vessel size does not change appreciably, this situation maybe approximated by a constant volume process. Using the thermodynamic diagram, the initial volume is about 0.20 ft^3/lb. At this same volume and final temperature, the pressure is:

P_{final} = 900 psia

Example 2 - Reactor Size

A batch reactor is to contain 1,000 lb of chlorine (Cl2) at 200 psia and 400 F. Estimate the reactor size.

Using the thermodynamic diagram, the volume is about 0.61 ft^3/lb at these conditions. Substitution of this into the equation below for the reactor size provides:

Reactor Size = (1,000 lb) (0.61 ft^3/lb) = 610 ft^3

Example 3 - Process Vessel Size

A process vessel is to contain 300 lb of at 300 psia and 600 F. Estimate the process vessel size.

Using the thermodynamic diagram, the volume is about 0.51 ft^3/lb at these conditions. Substitution of this into the equation below for the process vessel size provides:

Vessel Size = (300 lb) (0.51 ft^3/lb) = 153 ft^3

Example 4 - Heat Exchanger Duty

Chlorine (Cl2, 30,000 lb/hr) at 100 psia and 200 F is heated to 800 F and then fed to a plug-flow reactor. Estimate the heat exchanger duty necessary to accomplish the heating.

Substitution of mass flow and enthalpies from the thermodynamic diagram into the equation below provides:

Heat Exchanger Duty = mass flow (H_2 - H_1) = (30,000 lb/hr)(85 - 10) BTU/lb

= 2.25 million BTU/hr

Example 5 - Compression

Chlorine (Cl2, 20,000 lb/hr) at 10 psia and 0 F is compressed to 700 psia. Estimate the change in enthalpy for the compression assuming adibatic and reversible conditions (constant entropy).

Substitution of mass flow and enthalpies from the thermodynamic diagram into the equation below provides:

Enthalpy Change = mass flow (H$_2$ - H$_1$) = (20,000 lb/hr)(80 - (-10)) BTU/lb

= 1.8 million BTU/hr

This change in enthalpy represents energy that is required to accomplish the compression under adibatic and reversible conditions. Under operating conditions, the actual energy that is required for the compression will be somewhat more depending on the efficiency.

Example 6 - Expansion

Chlorine (Cl2, 30,000 lb/hr) at 600 psia and 750 F is expanded to 20 psia. Estimate the change in enthalpy for the expansion assuming adibatic and reversible conditions (constant entropy).

Substitution of mass flow and enthalpies from the thermodynamic diagram into the equation below provides:

Enthalpy Change = mass flow (H$_2$ - H$_1$) = (30,000 lb/hr)(0 - 75) BTU/lb

= - 2.25 million BTU/hr

This change in enthalpy represents energy that is available from the expansion under adibatic and reversible conditions. Under operating conditions, the actual energy that is available for the expansion will be somewhat less depending on the efficiency.

350

Appendix D

CRITICAL PROPERTIES AND ACENTRIC FACTOR FOR INORGANIC COMPOUNDS AND ELEMENTS*

Carl L. Yaws
Lamar University, Beaumont, Texas

NO	FORMULA	NAME	MW g/mol	T_F K	T_B K	T_C K	P_C bar	V_C cm³/mol	ρ_C g/cm³	Z_C	ω	SOURCE
1	Ag	SILVER	107.868	1234.00	2485.00	7480.00	5066.0	58.20	1.8534	0.474	0.150	1,6
2	AgCl	SILVER CHLORIDE	143.321	728.15	1837.15	---	---	---	---	---	---	2
3	AgI	SILVER IODIDE	234.773	825.15	1779.15	---	---	---	---	---	---	2
4	Al	ALUMINUM	26.982	933.00	2329.15	7151.00	5458.0	39.00	0.6918	0.358	---	1
5	AlB3H12	ALUMINUM BOROHYDRIDE	71.510	209.15	319.05	513.77	---	---	---	---	---	2,21
6	AlBr3	ALUMINUM BROMIDE	266.694	390.15	529.45	763.00	28.90	310.00	0.8603	0.141	0.399	3,6,10
7	AlCl3	ALUMINUM CHLORIDE	133.340	465.70	453.15	629.00	26.35	261.45	0.5100	0.132	0.660	1,10
8	AlF3	ALUMINUM FLUORIDE	83.977	1313.15	1810.15	---	---	---	---	---	---	2
9	AlI3	ALUMINUM IODIDE	407.695	464.15	658.65	983.00	---	408.00	0.9993	---	---	2,7,10
10	Al2O3	ALUMINUM OXIDE	101.961	2325.00	3253.15	5335.00	---	---	---	---	---	1
11	Al2S3O12	ALUMINUM SULFATE	342.154	1043.20	---	---	---	---	---	---	---	1
12	Ar	ARGON	39.948	83.80	87.28	150.86	48.98	74.59	0.5356	0.291	0.000	1
13	As	ARSENIC	74.922	1090.15	885.00	1673.15	223.00	34.90	2.1468	0.056	0.121	4,5,10
14	AsBr3	ARSENIC TRIBROMIDE	314.634	306.15	493.15	789.01	66.40	270.69	1.1623	0.274	---	2,7,21
15	AsCl3	ARSENIC TRICHLORIDE	181.280	255.15	403.55	654.00	59.12	252.00	0.7194	0.274	---	2,10,21
16	AsF3	ARSENIC TRIFLUORIDE	131.917	267.25	329.45	530.21	87.81	137.55	0.9590	0.274	---	2,21
17	AsF5	ARSENIC PENTAFLUORIDE	169.914	193.35	220.35	357.73	41.13	198.14	0.8575	0.274	---	2,21
18	AsH3	ARSINE	77.945	156.28	210.67	373.00	64.13	132.50	0.5883	0.274	0.006	1
19	AsI3	ARSENIC TRIIODIDE	455.635	419.15	676.15	---	---	---	---	---	---	2
20	As2O3	ARSENIC TRIOXIDE	197.841	585.95	730.35	---	---	---	---	---	---	3
21	At	ASTATINE	210.000	575.15	607.00	---	---	---	---	---	---	4,5
22	Au	GOLD	196.967	1337.33	3120.00	4398.00	---	50.30	3.9158	---	---	4,5,6
23	B	BORON	10.811	2348.15	4133.00	---	---	---	---	---	---	4,5
24	BBr3	BORON TRIBROMIDE	250.523	228.15	364.85	581.00	48.66	272.00	0.9210	0.274	---	2,10,21
25	BCl3	BORON TRICHLORIDE	117.169	166.15	285.65	451.95	38.71	265.99	0.4405	0.274	0.151	1
26	BF3	BORON TRIFLUORIDE	67.806	146.05	173.35	260.90	49.85	123.61	0.5485	0.284	0.430	1
27	BH2CO	BORINE CARBONYL	40.837	136.15	209.15	340.03	55.03	140.74	0.2902	0.274	---	3,21
28	BH3O3	BORIC ACID	61.833	458.15	---	---	---	---	---	---	---	1
29	B2D6	DEUTERODIBORANE	33.718	---	179.87	293.74	32.17	208.01	0.1621	0.274	---	2,21
30	B2H5Br	DIBORANE HYDROBROMIDE	106.566	168.95	289.45	466.98	43.61	243.93	0.4369	0.274	---	3,21
31	B2H6	DIBORANE	27.670	107.65	180.65	289.80	40.53	173.10	0.1598	0.291	0.125	1,14
32	B3N3H6	BORINE TRIAMINE	80.501	214.95	323.75	521.20	36.34	326.72	0.2464	0.274	---	2,21
33	B4H10	TETRABORANE	53.323	153.25	289.25	466.66	38.84	273.67	0.1948	0.274	---	2,21
34	B5H9	PENTABORANE	63.126	226.35	331.55	568.45	46.41	285.10	0.2214	0.280	---	2,8,14
35	B5H11	TETRAHYDROPENTABORANE	65.142	---	340.15	547.13	41.29	301.82	0.2158	0.274	---	2,21
36	B10H14	DECABORANE	122.221	372.75	486.15	---	---	---	---	---	---	2,4
37	Ba	BARIUM	137.327	1000.15	1907.00	---	---	---	---	---	---	2,4,5
38	Be	BERYLLIUM	9.012	1560.15	2744.00	---	---	---	---	---	---	4,5
39	BeB2H8	BERYLLIUM BOROHYDRIDE	38.698	396.15	363.15	---	---	---	---	---	---	2
40	BeBr2	BERYLLIUM BROMIDE	168.820	763.15	747.15	---	---	---	---	---	---	3
41	BeCl2	BERYLLIUM CHLORIDE	79.918	678.15	760.15	---	---	---	---	---	---	3
42	BeF2	BERYLLIUM FLUORIDE	47.009	1073.15	---	---	---	---	---	---	---	2,7
43	BeI2	BERYLLIUM IODIDE	262.821	761.15	760.15	---	---	---	---	---	---	3
44	Bi	BISMUTH	208.980	544.15	1698.15	4620.00	---	79.40	2.6320	---	---	3,6
45	BiBr3	BISMUTH TRIBROMIDE	448.692	491.15	734.15	1220.00	---	302.00	1.4857	---	---	2,6
46	BiCl3	BISMUTH TRICHLORIDE	315.338	503.15	714.15	1178.00	---	261.70	1.2050	---	---	2,6
47	BrF5	BROMINE PENTAFLUORIDE	174.896	211.75	313.55	470.00	57.16	187.31	0.9337	0.274	---	2,6,21
48	Br2	BROMINE	159.808	265.90	331.90	584.15	103.35	135.00	1.1838	0.287	0.119	1
49	C	CARBON	12.011	4247.00	4203.00	6810.00	2230.0	18.80	0.6389	0.074	1.566	1
50	CCl2O	PHOSGENE	98.916	145.37	280.71	455.00	56.74	190.22	0.5200	0.285	0.201	1
51	CF2O	CARBONYL FLUORIDE	66.007	161.89	188.58	297.00	57.60	141.00	0.4681	0.329	0.283	1
52	CH4N2O	UREA	60.056	405.85	465.00	705.00	90.50	218.00	0.2755	0.337	---	1
53	CH4N2S	THIOUREA	76.122	454.15	536.00	854.00	82.30	248.00	0.3069	0.287	0.359	1
54	CNBr	CYANOGEN BROMIDE	105.922	331.15	334.65	---	---	---	---	---	---	2
55	CNCl	CYANOGEN CHLORIDE	61.470	266.65	286.00	449.00	59.90	163.00	0.3771	0.262	0.320	1
56	CNF	CYANOGEN FLUORIDE	45.016	---	227.17	368.51	79.00	106.26	0.4236	0.274	---	2,21
57	CO	CARBON MONOXIDE	28.010	68.15	81.70	132.92	34.99	93.10	0.3009	0.295	0.066	1
58	COS	CARBONYL SULFIDE	60.076	134.35	223.00	378.80	63.49	135.10	0.4447	0.272	0.097	1

* A computer program, containing data for all compounds, is available for a nominal fee (Carl L. Yaws, Box 10053, Lamar University, Beaumont, TX 77710, phone/FAX 409-880-8787). The computer program is in ASCII which can be accessed by other software.

NO	FORMULA	NAME	MW g/mol	T_F K	T_B K	T_C K	P_C bar	V_C cm³/mol	ρ_C g/cm³	Z_C	ω	SOURCE
59	COSe	CARBON OXYSELENIDE	106.970	---	251.25	406.58	86.30	107.32	0.9967	0.274	---	3,21
60	CO2	CARBON DIOXIDE	44.010	216.58	194.70	304.19	73.81	94.00	0.4682	0.274	0.228	1
61	CS2	CARBON DISULFIDE	76.143	161.58	319.37	552.00	79.03	160.00	0.4759	0.276	0.108	1
62	CSeS	CARBON SELENOSULFIDE	123.037	197.95	358.75	576.53	74.12	177.18	0.6944	0.274	---	3,21
63	C2N2	CYANOGEN	52.035	238.75	252.15	399.90	63.03	144.52	0.3601	0.274	---	3,6
64	C3S2	CARBON SUBSULFIDE	100.165	273.55	---	---	---	---	---	---	---	3
65	Ca	CALCIUM	40.078	1115.15	1762.00	---	---	---	---	---	---	4,5
66	CaF2	CALCIUM FLUORIDE	78.075	1691.00	2806.50	---	---	---	---	---	---	1
67	CbF5	COLUMBIUM FLUORIDE	187.898	348.65	498.15	---	---	---	---	---	---	3
68	Cd	CADMIUM	112.411	594.05	1043.15	2291.00	---	37.90	2.9660	---	---	3,6
69	CdCl2	CADMIUM CHLORIDE	183.316	841.15	1240.15	---	---	---	---	---	---	3
70	CdF2	CADMIUM FLUORIDE	150.408	793.15	2024.15	---	---	---	---	---	---	3
71	CdI2	CADMIUM IODIDE	366.220	658.15	1069.15	---	---	---	---	---	---	3
72	CdO	CADMIUM OXIDE	128.410	---	1832.15	---	---	---	---	---	---	3
73	ClF	CHLORINE MONOFLUORIDE	54.451	128.15	172.65	282.32	79.01	81.39	0.6690	0.274	---	2,21
74	ClFO3	PERCHLORYL FLUORIDE	102.449	125.41	226.49	368.40	53.70	161.00	0.6363	0.282	0.173	1
75	ClF3	CHLORINE TRIFLUORIDE	92.448	190.15	284.65	459.39	77.79	134.52	0.6872	0.274	---	2
76	ClF5	CHLORINE PENTAFLUORIDE	130.445	---	260.05	415.90	52.60	230.40	0.5662	0.350	0.216	2,6,10
77	ClHO3S	CHLOROSULFONIC ACID	116.525	193.15	427.00	700.00	85.00	195.00	0.5976	0.285	0.301	1
78	ClHO4	PERCHLORIC ACID	100.458	171.95	385.00	631.00	38.60	168.00	0.5980	0.124	0.050	1
79	ClO2	CHLORINE DIOXIDE	67.452	213.55	284.05	465.00	108.28	97.83	0.6895	0.274	0.356	1
80	Cl2	CHLORINE	70.905	172.12	239.12	417.15	77.11	123.75	0.5730	0.275	0.069	1
81	Cl2O	CHLORINE MONOXIDE	86.905	157.15	275.35	444.68	74.94	135.16	0.6430	0.274	---	2,21
82	Cl2O7	CHLORINE HEPTOXIDE	182.901	182.15	351.95	565.78	50.90	253.23	0.7223	0.274	---	2,21
83	Co	COBALT	58.933	1768.15	2528.00	---	---	---	---	---	---	4,5
84	CoCl2	COBALT CHLORIDE	129.839	1008.15	1323.15	---	---	---	---	---	---	2
85	CoNC3O4	COBALT NITROSYL TRICARBONYL	172.971	262.15	353.15	567.68	---	---	---	---	---	3,21
86	Cr	CHROMIUM	51.996	2180.15	2840.00	---	---	---	---	---	---	2,4,5
87	CrC6O6	CHROMIUM CARBONYL	220.059	423.65	424.15	---	---	---	---	---	---	3,7
88	CrO2Cl2	CHROMIUM OXYCHLORIDE	154.900	176.65	390.25	626.33	59.99	237.81	0.6513	0.274	---	2,7,21
89	Cs	CESIUM	132.905	301.65	963.15	2048.10	116.50	316.40	0.4201	0.216	---	2,6
90	CsBr	CESIUM BROMIDE	212.809	909.15	1573.15	---	---	---	---	---	---	3
91	CsCl	CESIUM CHLORIDE	168.358	919.15	1573.15	---	---	---	---	---	---	3
92	CsF	CESIUM FLUORIDE	151.904	956.15	1524.15	---	---	---	---	---	---	3
93	CsI	CESIUM IODIDE	259.810	894.15	1553.15	---	---	---	---	---	---	3
94	Cu	COPPER	63.546	1357.77	3150.00	5123.00	---	61.00	1.0417	---	---	4,5,6
95	CuBr	CUPROUS BROMIDE	143.450	777.15	1628.15	---	---	---	---	---	---	2
96	CuCl	CUPROUS CHLORIDE	98.999	703.00	1763.15	2435.00	---	---	---	---	---	1
97	CuCl2	CUPRIC CHLORIDE	134.451	906.15	1266.15	2010.00	---	---	---	---	---	1
98	CuI	COPPER IODIDE	190.450	878.15	1609.15	---	---	---	---	---	---	3
99	DCN	DEUTERIUM CYANIDE	28.034	261.15	299.35	482.63	113.56	96.81	0.2896	0.274	---	2,21
100	D2	DEUTERIUM	4.032	18.73	23.65	38.35	16.64	60.26	0.0669	0.314	-0.14	1
101	D2O	DEUTERIUM OXIDE	20.031	276.96	374.55	643.89	219.41	56.30	0.3558	0.231	0.368	1
102	Eu	EUROPIUM	151.965	1095.15	1742.00	5150.00	---	---	---	---	---	4,5,6
103	F2	FLUORINE	37.997	53.53	84.95	144.31	52.15	66.20	0.5740	0.288	0.059	1
104	F2O	FLUORINE OXIDE	53.996	49.25	128.55	215.10	49.50	97.60	0.5532	0.270	---	2,6
105	Fe	IRON	55.847	1808.15	3000.00	9340.00	10150	28.00	1.9945	0.366	-0.30	1
106	FeC5O5	IRON PENTACARBONYL	195.899	252.15	378.15	607.20	35.24	392.52	0.4991	0.274	---	3,21
107	FeCl2	FERROUS CHLORIDE	126.752	945.15	1299.15	---	---	---	---	---	---	3,7
108	FeCl3	FERRIC CHLORIDE	162.205	577.15	592.15	---	---	---	---	---	---	3
109	Fr	FRANCIUM	223.000	300.15	879.00	---	---	---	---	---	---	4,5
110	Ga	GALLIUM	69.723	302.91	2517.00	7620.00	---	75.30	0.9259	---	---	4,5,6
111	GaCl3	GALLIUM TRICHLORIDE	176.081	350.90	474.15	694.00	38.20	263.00	0.6695	0.174	0.458	1
112	Gd	GADOLINIUM	157.250	1587.15	1770.00	---	---	---	---	---	---	4,5
113	Ge	GERMANIUM	72.610	1211.40	3125.00	8400.00	---	---	---	---	---	4,5,6
114	GeBr4	GERMANIUM BROMIDE	392.226	299.25	462.15	740.00	44.17	381.66	1.0277	0.274	---	2,21
115	GeCl4	GERMANIUM CHLORIDE	214.421	223.65	357.15	574.00	39.57	330.45	0.6489	0.274	---	2,21
116	GeHCl3	TRICHLORO GERMANE	179.976	202.05	348.15	559.78	46.21	275.96	0.6522	0.274	---	2,21
117	GeH4	GERMANE	76.642	107.26	185.00	308.00	55.50	140.00	0.5474	0.303	0.151	1
118	Ge2H6	DIGERMANE	151.268	164.15	304.65	491.01	46.67	239.68	0.6311	0.274	---	2,21
119	Ge3H8	TRIGERMANE	225.894	167.55	383.95	616.37	47.05	298.44	0.7569	0.274	---	2,21
120	HBr	HYDROGEN BROMIDE	80.912	186.34	206.45	363.15	85.52	100.26	0.8070	0.284	0.069	1
121	HCN	HYDROGEN CYANIDE	27.026	259.91	298.85	456.65	53.91	138.59	0.1950	0.197	0.410	1
122	HCl	HYDROGEN CHLORIDE	36.461	158.97	188.15	324.65	83.09	81.02	0.4500	0.249	0.132	1
123	HF	HYDROGEN FLUORIDE	20.006	189.79	292.67	461.15	64.85	69.00	0.2899	0.117	0.383	1
124	HI	HYDROGEN IODIDE	127.912	222.38	237.55	423.85	83.10	121.94	1.0490	0.288	0.038	1
125	HNO3	NITRIC ACID	63.013	231.55	356.15	520.00	68.90	145.00	0.4346	0.231	0.714	1
126	H2	HYDROGEN	2.016	13.95	20.39	33.18	13.13	64.15	0.0314	0.305	-0.22	1
127	H2O	WATER	18.015	273.15	373.15	647.13	220.55	55.95	0.3220	0.229	0.345	1

* A computer program, containing data for all compounds, is available for a nominal fee (Carl L. Yaws, Box 10053, Lamar University, Beaumont, TX 77710, phone/FAX 409-880-8787). The computer program is in ASCII which can be accessed by other software.

NO	FORMULA	NAME	MW g/mol	T_F K	T_B K	T_C K	P_C bar	V_C cm³/mol	ρ_C g/cm³	Z_C	ω	SOURCE
128	H2O2	HYDROGEN PEROXIDE	34.015	272.72	423.35	730.15	216.84	77.70	0.4378	0.278	0.360	1
129	H2S	HYDROGEN SULFIDE	34.082	187.68	212.80	373.53	89.63	98.49	0.3460	0.284	0.083	1
130	H2SO4	SULFURIC ACID	98.079	283.46	610.00	925.00	64.00	177.03	0.5540	0.147	---	1
131	H2S2	HYDROGEN DISULFIDE	66.148	183.45	337.15	542.39	88.36	139.83	0.4731	0.274	---	2,21
132	H2Se	HYDROGEN SELENIDE	80.976	209.15	232.05	411.10	83.44	112.24	0.7215	0.274	---	2,6
133	H2Te	HYDROGEN TELLURIDE	129.616	224.15	271.15	438.04	71.93	138.73	0.9343	0.274	---	2,21
134	H3NO3S	SULFAMIC ACID	97.095	478.00	---	---	---	225.00	0.4315	---	---	1
135	He	HELIUM-3	3.016	1.01	3.20	3.31	1.17	72.50	0.0416	0.308	-0.47	1
136	He	HELIUM-4	4.003	1.76	4.22	5.20	2.28	57.30	0.0699	0.302	-0.39	1
137	Hf	HAFNIUM	178.490	2506.15	5960.00	---	---	---	---	---	---	4,5
138	Hg	MERCURY	200.590	234.29	629.73	1735.00	1608.0	56.35	3.5597	0.628	-0.16	1
139	HgBr2	MERCURIC BROMIDE	360.398	510.15	592.15	---	---	---	---	---	---	3
140	HgCl2	MERCURIC CHLORIDE	271.495	550.15	577.15	---	---	---	---	---	---	3
141	HgI2	MERCURIC IODIDE	454.399	532.15	627.15	1078.10	100.00	---	---	---	---	3,6
142	IF7	IODINE HEPTAFLUORIDE	259.893	278.65	277.15	447.53	41.26	247.10	1.0518	0.274	---	2,21
143	I2	IODINE	253.809	386.75	458.39	819.15	116.54	155.00	1.6375	0.265	0.117	1
144	In	INDIUM	114.818	429.75	2323.00	6730.00	2432.0	82.60	1.3900	0.359	---	4,5,6
145	Ir	IRIDIUM	192.220	2719.15	4450.00	---	---	---	---	---	---	4,5
146	K	POTASSIUM	39.098	336.35	1037.00	2223.00	162.12	209.00	0.1871	0.183	-0.18	1
147	KBr	POTASSIUM BROMIDE	119.002	1003.15	1656.15	---	---	---	---	---	---	2
148	KCl	POTASSIUM CHLORIDE	74.551	1044.00	1688.87	3470.00	180.00	625.00	0.1193	0.39	-0.12	1
149	KF	POTASSIUM FLUORIDE	58.097	1153.15	1775.15	---	---	---	---	---	---	2
150	KI	POTASSIUM IODIDE	166.003	996.15	1597.15	---	---	---	---	---	---	2
151	KOH	POTASSIUM HYDROXIDE	56.106	679.00	1600.00	---	---	---	---	---	---	1
152	Kr	KRYPTON	83.800	115.78	119.80	209.35	55.02	91.20	0.9189	0.288	0.000	1
153	La	LANTHANUM	138.906	1193.15	3643.00	9511.00	5460.0	36.50	3.8056	0.252	---	4,5,6
154	Li	LITHIUM	6.941	453.69	1597.00	4085.00	1722.5	47.00	0.1477	0.238	-0.04	1
155	LiBr	LITHIUM BROMIDE	86.845	820.15	1583.15	---	---	---	---	---	---	2
156	LiCl	LITHIUM CHLORIDE	42.394	887.15	1655.15	---	---	---	---	---	---	3
157	LiF	LITHIUM FLUORIDE	25.939	1143.15	1954.15	---	---	---	---	---	---	2
158	LiI	LITHIUM IODIDE	133.845	719.15	1444.15	---	---	---	---	---	---	2
159	Lu	LUTECIUM	174.967	1936.15	2535.00	---	---	---	---	---	---	4,5
160	Mg	MAGNESIUM	24.305	923.15	1376.00	---	---	---	---	---	---	4,5
161	MgCl2	MAGNESIUM CHLORIDE	95.210	985.15	1691.15	---	---	---	---	---	---	3
162	MgO	MAGNESIUM OXIDE	40.304	3105.00	3873.20	5950.00	33.91	209.50	0.1924	0.014	0.214	1
163	Mn	MANGANESE	54.938	1519.15	2392.00	---	---	---	---	---	---	4,5
164	MnCl2	MANGANESE CHLORIDE	125.843	923.15	1463.15	---	---	---	---	---	---	3
165	Mo	MOLYBDENUM	95.940	2895.15	5081.15	9620.00	---	38.30	2.5050	---	---	3,6
166	MoF6	MOLYBDENUM FLUORIDE	209.930	290.15	309.15	498.12	50.30	225.58	0.9306	0.274	---	2,21
167	MoO3	MOLYBDENUM OXIDE	143.938	1068.15	1424.15	---	---	---	---	---	---	2
168	NCl3	NITROGEN TRICHLORIDE	120.365	246.15	344.15	564.00	62.10	206.90	0.5818	0.274	---	1
169	ND3	HEAVY AMMONIA	20.055	199.15	239.75	388.40	125.71	70.38	0.2850	0.274	---	2,21
170	NF3	NITROGEN TRIFLUORIDE	71.002	66.36	144.09	233.85	45.30	118.75	0.5979	0.277	0.126	1
171	NH3	AMMONIA	17.031	195.41	239.72	405.65	112.78	72.47	0.2350	0.242	0.252	1
172	NH3O	HYDROXYLAMINE	33.030	306.25	383.00	574.00	175.18	74.64	0.4425	0.274	0.694	1
173	NH4Br	AMMONIUM BROMIDE	97.943	---	669.15	---	---	---	---	---	---	2
174	NH4Cl	AMMONIUM CHLORIDE	53.491	793.20	612.00	882.00	16.40	---	---	---	3.920	1,10
175	NH4I	AMMONIUM IODIDE	144.943	---	678.05	---	---	---	---	---	---	2
176	NH5O	AMMONIUM HYDROXIDE	35.046	194.15	---	---	---	---	---	---	---	1
177	NH5S	AMMONIUM HYDROGENSULFIDE	51.112	391.15	306.45	---	---	---	---	---	---	2,5
178	NO	NITRIC OXIDE	30.006	112.15	121.38	180.15	64.85	57.70	0.5200	0.250	0.585	1
179	NOCl	NITROSYL CHLORIDE	65.459	213.55	267.77	440.65	91.19	139.30	0.4699	0.347	0.307	1
180	NOF	NITROSYL FLUORIDE	49.005	139.15	217.15	352.67	112.78	71.24	0.6879	0.274	---	3,21
181	NO2	NITROGEN DIOXIDE	46.006	261.95	294.00	431.35	101.33	82.49	0.5577	0.233	0.849	1
182	N2	NITROGEN	28.013	63.15	77.35	126.10	33.94	90.10	0.3109	0.292	0.040	1
183	N2F4	TETRAFLUOROHYDRAZINE	104.007	111.65	198.95	309.35	37.10	213.00	0.4883	0.307	0.223	1
184	N2H4	HYDRAZINE	32.045	274.69	386.65	653.15	146.92	158.00	0.2028	0.427	0.314	1
185	N2H4C	AMMONIUM CYANIDE	44.056	309.15	304.85	491.32	109.47	102.24	0.4309	0.274	---	2,21
186	N2H6CO2	AMMONIUM CARBAMATE	78.071	---	331.45	---	---	---	---	---	---	2
187	N2O	NITROUS OXIDE	44.013	182.33	184.67	309.57	72.45	97.37	0.4520	0.274	0.142	1
188	N2O3	NITROGEN TRIOXIDE	76.012	170.00	275.15	425.00	69.90	195.00	0.3898	0.386	0.431	1
189	N2O4	NITROGEN TETRAOXIDE	92.011	261.90	302.22	431.15	101.33	82.49	1.1154	0.233	1.007	1
190	N2O5	NITROGEN PENTOXIDE	108.010	303.15	320.15	515.51	64.33	182.56	0.5917	0.274	---	1,9,21
191	Na	SODIUM	22.990	370.98	1156.00	2573.00	354.64	116.00	0.1982	0.192	-0.10	1
192	NaBr	SODIUM BROMIDE	102.894	1020.00	1663.82	4287.00	192.52	398.00	0.2585	0.215	-0.80	1
193	NaCN	SODIUM CYANIDE	49.008	836.85	1769.15	2900.00	---	---	---	---	---	1
194	NaCl	SODIUM CHLORIDE	58.442	1073.95	1738.15	3400.00	355.00	266.00	0.2197	0.334	0.134	1
195	NaF	SODIUM FLUORIDE	41.988	1269.00	1982.72	5530.00	531.96	185.00	0.2270	0.214	-1.11	1
196	NaI	SODIUM IODIDE	149.894	924.15	1577.15	---	---	---	---	---	---	2

* A computer program, containing data for all compounds, is available for a nominal fee (Carl L. Yaws, Box 10053, Lamar University, Beaumont, TX 77710, phone/FAX 409-880-8787). The computer program is in ASCII which can be accessed by other software.

NO	FORMULA	NAME	MW g/mol	T_F K	T_B K	T_C K	P_C bar	V_C cm³/mol	ρ_C g/cm³	Z_C	ω	SOURCE
197	NaOH	SODIUM HYDROXIDE	39.997	596.00	1663.15	2820.00	253.31	200.00	0.2000	0.216	---	1,7
198	Na2SO4	SODIUM SULFATE	142.043	1157.00	---	---	---	---	---	---	---	1
199	Nb	NIOBIUM	92.906	2750.15	5115.00	---	---	---	---	---	---	4,5
200	Nd	NEODYMIUM	144.240	1289.15	3384.00	---	---	---	---	---	---	4,5
201	Ne	NEON	20.180	24.55	27.09	44.40	26.53	41.70	0.4839	0.300	-0.04	1
202	Ni	NICKEL	58.693	1728.15	2415.00	---	---	---	---	---	---	4,5
203	NiC4O4	NICKEL CARBONYL	170.735	248.15	315.65	508.40	32.39	357.53	0.4775	0.274	---	2,21
204	NiF2	NICKEL FLUORIDE	96.690	1723.15	2013.15	---	---	---	---	---	---	2,9
205	Np	NEPTUNIUM	237.000	913.15	---	---	---	---	---	---	---	2,7
206	O2	OXYGEN	31.999	54.36	90.17	154.58	50.43	73.40	0.4360	0.288	0.022	1
207	O3	OZONE	47.998	80.15	161.85	261.00	55.73	89.00	0.5393	0.229	0.227	1
208	Os	OSMIUM	190.230	3306.15	4880.00	---	---	---	---	---	---	4,5
209	OsOF5	OSMIUM OXIDE PENTAFLUORIDE	301.221	332.95	373.65	---	---	---	---	---	---	2,9
210	OsO4	OSMIUM TETROXIDE - YELLOW	254.228	329.15	403.15	---	---	---	---	---	---	3
211	OsO4	OSMIUM TETROXIDE - WHITE	254.228	315.15	403.15	---	---	---	---	---	---	3
212	P	PHOSPHORUS - WHITE	30.974	317.25	553.45	993.75	83.29	---	---	---	---	1
213	PBr3	PHOSPHORUS TRIBROMIDE	270.686	233.15	448.45	711.00	53.99	300.00	0.9023	0.274	---	2,10,21
214	PCl2F3	PHOSPHORUS DICHLORIDE TRIFLUORIDE	158.874	265.15	283.15	457.02	40.48	257.17	0.6178	0.274	---	2,5,21
215	PCl3	PHOSPHORUS TRICHLORIDE	137.332	181.15	349.25	563.15	56.70	260.00	0.5282	0.315	0.234	1
216	PCl5	PHOSPHORUS PENTACHLORIDE	208.237	433.15	433.00	646.15	---	---	---	---	---	1
217	PH3	PHOSPHINE	33.998	139.37	185.41	324.75	65.36	113.32	0.3000	0.274	0.036	1
218	PH4Br	PHOSPHONIUM BROMIDE	114.910	---	311.45	501.76	62.26	183.57	0.6260	0.274	---	2,21
219	PH4Cl	PHOSPHONIUM CHLORIDE	70.458	244.65	246.15	322.30	49.14	149.42	0.4716	0.274	1.64	2,10
220	PH4I	PHOSPHONIUM IODIDE	161.910	291.65	335.45	539.70	77.61	158.40	1.0221	0.274	---	2,7,21
221	POCl3	PHOSPHORUS OXYCHLORIDE	153.331	274.33	378.65	602.15	51.66	265.54	0.5774	0.274	---	1,21
222	PSBr3	PHOSPHORUS THIOBROMIDE	302.752	311.15	448.15	---	---	---	---	---	---	2
223	PSCl3	PHOSPHORUS THIOCHLORIDE	169.398	236.95	398.15	638.82	48.57	299.58	0.5654	0.274	---	12,21
224	P4O6	PHOSPHORUS TRIOXIDE	219.891	295.65	446.25	714.86	52.08	312.69	0.7032	0.274	---	2,21
225	P4O10	PHOSPHORUS PENTOXIDE	283.889	693.15	---	---	---	---	---	---	---	1
226	P4S10	PHOSPHORUS PENTASULFIDE	444.555	561.15	787.15	1291.00	232.00	---	---	---	0.594	2
227	Pb	LEAD	207.200	600.61	2024.00	5400.00	861.30	93.20	2.2232	0.179	---	4,5,6
228	PbBr2	LEAD BROMIDE	367.008	646.15	1187.15	---	---	---	---	---	---	3
229	PbCl2	LEAD CHLORIDE	278.105	774.15	1227.15	---	---	---	---	---	---	2
230	PbF2	LEAD FLUORIDE	245.197	1128.15	1566.15	---	---	---	---	---	---	3
231	PbI2	LEAD IODIDE	461.009	675.15	1145.15	---	---	---	---	---	---	3
232	PbO	LEAD OXIDE	223.199	1163.15	1745.15	---	---	---	---	---	---	3
233	PbS	LEAD SULFIDE	239.266	1387.15	1554.15	---	---	---	---	---	---	3
234	Pd	PALLADIUM	106.420	1828.05	3385.00	---	---	---	---	---	---	4,5
235	Po	POLONIUM	209.000	527.15	1235.00	---	---	---	---	---	---	4,5
236	Pt	PLATINUM	195.080	2041.55	3980.00	6983.00	---	759.10	0.2570	---	---	4,5,6
237	Ra	RADIUM	226.000	973.15	1809.00	---	---	---	---	---	---	4,5
238	Rb	RUBIDIUM	85.468	312.46	978.00	2111.10	134.00	247.00	0.3460	0.189	---	4,5,6
239	RbBr	RUBIDIUM BROMIDE	165.372	955.15	1625.15	---	---	---	---	---	---	2
240	RbCl	RUBIDIUM CHLORIDE	120.921	988.15	1654.15	---	---	---	---	---	---	2
241	RbF	RUBIDIUM FLUORIDE	104.466	1033.15	1681.15	---	---	---	---	---	---	2
242	RbI	RUBIDIUM IODIDE	212.372	915.15	1577.15	---	---	---	---	---	---	2
243	Re	RHENIUM	186.207	3459.15	5915.00	---	---	32.10	5.8008	---	---	4,5,6
244	Re2O7	RHENIUM HEPTOXIDE	484.410	569.15	635.55	---	---	---	---	---	---	3
245	Rh	RHODIUM	102.906	2237.15	3940.00	---	---	---	---	---	---	4,5
246	Rn	RADON	222.000	202.15	211.35	377.40	63.00	140.00	1.5857	0.281	---	2,6
247	Ru	RUTHENIUM	101.070	2607.15	4500.00	---	---	---	---	---	---	4,5
248	RuF5	RUTHENIUM PENTAFLUORIDE	196.062	359.65	600.15	---	---	---	---	---	---	2,9
249	S	SULFUR	32.066	388.36	717.82	1313.00	182.08	158.00	0.2029	0.264	0.262	1
250	SF4	SULFUR TETRAFLUORIDE	108.060	149.15	233.15	364.00	52.22	158.77	0.6806	0.274	---	2,7,10,2
251	SF6	SULFUR HEXAFLUORIDE	146.056	222.45	209.25	318.69	37.60	198.52	0.7357	0.282	0.215	1,15
252	SOBr2	THIONYL BROMIDE	207.873	220.95	412.65	661.75	64.89	232.32	0.8948	0.274	---	2,21
253	SOCl2	THIONYL CHLORIDE	118.971	172.00	348.75	567.00	63.63	203.00	0.5861	0.274	---	1,21
254	SOF2	SULFUROUS OXYFLUORIDE	86.062	162.65	228.90	371.25	59.28	142.65	0.6033	0.274	---	2,21
255	SO2	SULFUR DIOXIDE	64.065	200.00	263.13	430.75	78.84	122.00	0.5251	0.269	0.245	1
256	SO2Cl2	SULFURYL CHLORIDE	134.970	222.00	342.55	545.00	46.10	224.00	0.6025	0.228	0.176	1
257	SO3	SULFUR TRIOXIDE	80.064	289.95	317.90	490.85	82.07	127.08	0.6300	0.256	0.422	1
258	S2Cl2	SULFUR MONOCHLORIDE	135.037	193.15	411.15	659.37	62.75	239.38	0.5641	0.274	---	2,21
259	Sb	ANTIMONY	121.757	903.78	1898.00	5070.00	---	---	---	---	---	4,5,6
260	SbBr3	ANTIMONY TRIBROMIDE	361.469	369.75	548.15	---	---	---	---	---	---	3
261	SbCl3	ANTIMONY TRICHLORIDE	228.115	346.55	493.40	794.00	48.20	270.00	0.8449	0.197	0.171	1
262	SbCl5	ANTIMONY PENTACHLORIDE	299.021	275.95	413.15	662.54	39.42	382.86	0.7810	0.274	---	2,8,21
263	SbH3	STIBINE	124.781	185.15	255.15	440.35	73.06	157.20	0.7938	0.314	---	7,21
264	SbI3	ANTIMONY TRIIODIDE	502.470	440.15	674.15	---	---	---	---	---	---	3
265	Sb2O3	ANTIMONY TRIOXIDE	291.512	929.15	1698.15	---	---	---	---	---	---	3

* A computer program, containing data for all compounds, is available for a nominal fee (Carl L. Yaws, Box 10053, Lamar University, Beaumont, TX 77710, phone/FAX 409-880-8787). The computer program is in ASCII which can be accessed by other software.

NO	FORMULA	NAME	MW g/mol	T_F K	T_B K	T_C K	P_C bar	V_C cm³/mol	ρ_C g/cm³	Z_C	ω	SOURCE
266	Sc	SCANDIUM	44.956	1814.15	2700.00	---	---	---	---	---	---	4,5
267	Se	SELENIUM	78.960	494.15	930.00	1766.00	380.00	62.30	1.2674	0.161	---	4,5,6
268	SeCl4	SELENIUM TETRACHLORIDE	220.771	---	464.65	743.95	61.05	277.61	0.7953	0.274	---	2,21
269	SeF6	SELENIUM HEXAFLUORIDE	192.950	238.45	227.35	368.80	44.75	187.74	1.0278	0.274	---	2,21
270	SeOCl2	SELENIUM OXYCHLORIDE	165.865	281.65	441.15	706.80	77.47	207.84	0.7981	0.274	---	2,21
271	SeO2	SELENIUM DIOXIDE	110.959	613.15	590.15	---	---	---	---	---	---	2
272	Si	SILICON	28.086	1685.00	3513.80	5159.00	537.00	233.00	0.1205	0.292	---	1,17
273	SiBrCl2F	BROMODICHLOROFLUOROSILANE	197.893	160.85	308.55	497.17	38.27	295.93	0.6687	0.274	---	3,21
274	SiBrF3	TRIFLUOROBROMOSILANE	164.985	202.65	231.45	375.28	37.54	227.73	0.7245	0.274	---	2,21
275	SiBr2ClF	DIBROMOCHLOROFLUOROSILANE	242.345	173.85	332.65	535.27	39.32	310.14	0.7814	0.274	---	2,21
276	SiClF3	TRIFLUOROCHLOROSILANE	120.533	131.15	203.15	330.54	35.25	213.61	0.5643	0.274	---	2,21
277	SiCl2F2	DICHLORODIFLUOROSILANE	136.988	133.45	241.35	390.93	35.95	247.69	0.5531	0.274	---	2,21
278	SiCl3F	TRICHLOROFLUOROSILANE	153.442	152.35	285.35	460.49	37.24	281.72	0.5447	0.274	---	2,21
279	SiCl4	SILICON TETRACHLORIDE	169.896	204.30	330.00	507.00	35.93	326.00	0.5212	0.278	0.232	1
280	SiF4	SILICON TETRAFLUORIDE	104.079	186.35	178.35	259.00	37.19	165.00	0.6308	0.285	0.385	1,18
281	SiHBr3	TRIBROMOSILANE	268.805	199.65	384.95	610.00	47.02	350.00	0.7680	0.324	---	2,11
282	SiHCl3	TRICHLOROSILANE	135.452	144.95	305.00	479.00	41.70	268.00	0.5054	0.281	0.203	1,19
283	SiHF3	TRIFLUOROSILANE	86.089	141.75	178.15	291.02	39.95	165.93	0.5188	0.274	---	2,21
284	SiH2Br2	DIBROMOSILANE	189.909	202.95	343.65	550.00	53.00	246.00	0.7720	0.285	---	2,11
285	SiH2Cl2	DICHLOROSILANE	101.007	151.15	281.45	449.00	44.30	228.00	0.4430	0.271	0.177	1,20
286	SiH2F2	DIFLUOROSILANE	68.098	---	195.35	318.21	47.59	152.31	0.4471	0.274	---	2,21
287	SiH2I2	DIIODOSILANE	283.910	272.15	422.65	660.00	66.88	232.00	1.2238	0.283	---	2,11
288	SiH3Br	MONOBROMOSILANE	111.013	179.25	275.55	454.00	56.44	177.00	0.6272	0.265	---	2,11
289	SiH3Cl	MONOCHLOROSILANE	66.562	155.05	242.75	396.65	48.43	174.00	0.3825	0.256	0.136	3,12
290	SiH3F	MONOFLUOROSILANE	50.108	---	175.15	286.28	46.88	139.12	0.3602	0.274	---	2,21
291	SiH3I	IODOSILANE	158.014	216.15	318.55	515.00	69.41	160.00	0.9876	0.259	---	2,11
292	SiH4	SILANE	32.117	88.15	161.00	269.70	48.43	132.70	0.2420	0.287	0.097	1,16
293	SiO2	SILICON DIOXIDE	60.084	1883.00	2503.20	---	---	---	---	---	---	1
294	Si2Cl6	HEXACHLORODISILANE	268.887	271.95	412.15	660.96	29.11	517.27	0.5198	0.274	---	2,21
295	Si2F6	HEXAFLUORODISILANE	170.161	254.55	254.25	411.33	30.16	310.64	0.5478	0.274	---	2,21
296	Si2H5Cl	DISILANYL CHLORIDE	96.663	---	314.70	506.89	41.58	277.73	0.3480	0.274	---	2,21
297	Si2H6	DISILANE	62.219	140.65	259.00	432.00	51.30	198.00	0.3142	0.283	0.102	1
298	Si2OCl3F3	TRICHLOROTRIFLUORODISILOXANE	235.524	---	315.89	508.78	26.77	433.00	0.5439	0.274	---	2,21
299	Si2OCl6	HEXACHLORODISILOXANE	284.887	239.95	408.75	655.58	27.90	535.21	0.5323	0.274	---	2,21
300	Si2OH6	DISILOXANE	78.218	128.95	257.75	416.86	36.19	262.41	0.2981	0.274	---	2,21
301	Si3Cl8	OCTACHLOROTRISILANE	367.878	---	484.55	775.41	24.70	714.99	0.5145	0.274	---	2,21
302	Si3H8	TRISILANE	92.320	155.95	326.25	525.15	33.70	354.94	0.2601	0.274	---	2,21
303	Si3H9N	TRISILAZANE	107.335	167.45	321.85	518.20	31.65	372.91	0.2878	0.274	---	2,21
304	Si4H10	TETRASILANE	122.421	179.55	373.15	599.30	29.68	460.01	0.2661	0.274	---	2,21
305	Sm	SAMARIUM	150.360	1345.15	1874.00	---	---	---	---	---	---	4,5
306	Sn	TIN	118.710	505.08	2995.00	7400.00	---	115.10	1.0314	---	---	4,5,6
307	SnBr4	STANNIC BROMIDE	438.326	304.15	477.85	764.82	43.43	401.12	1.0928	0.274	---	2,21
308	SnCl2	STANNOUS CHLORIDE	189.615	519.95	896.15	---	---	---	---	---	---	3
309	SnCl4	STANNIC CHLORIDE	260.521	242.95	386.15	619.85	41.24	342.37	0.7609	0.274	---	2,21
310	SnH4	STANNIC HYDRIDE	122.742	123.25	220.85	358.52	53.42	152.88	0.8028	0.274	---	2,21
311	SnI4	STANNIC IODIDE	626.328	417.65	621.15	---	---	---	---	---	---	2
312	Sr	STRONTIUM	87.620	1050.15	1630.00	---	---	---	---	---	---	4,5
313	SrO	STRONTIUM OXIDE	103.619	2703.15	---	---	---	---	---	---	---	2
314	Ta	TANTALUM	180.948	3290.15	5565.00	---	---	---	---	---	---	4,5
315	Tc	TECNNETIUM	98.000	2430.15	5000.00	---	---	---	---	---	---	4,5
316	Te	TELLURIUM	127.600	722.66	1285.00	4840.00	---	---	---	---	---	4,5,6
317	TeCl4	TELLURIUM TETRACHLORIDE	269.411	497.15	665.15	---	---	---	---	---	---	2
318	TeF6	TELLURIUM HEXAFLUORIDE	241.590	235.35	234.55	380.18	34.47	251.23	0.9616	0.274	---	2,21
319	Ti	TITANIUM	47.880	1941.15	3442.00	6400.00	---	---	---	---	---	4,5,6
320	TiCl4	TITANIUM TETRACHLORIDE	189.691	249.05	409.00	638.00	46.61	340.00	0.5579	0.299	0.284	1
321	Tl	THALLIUM	204.383	577.15	1745.00	---	---	---	---	---	---	4,5
322	TlBr	THALLOUS BROMIDE	284.287	733.15	1092.15	---	---	---	---	---	---	2
323	TlI	THALLOUS IODIDE	331.288	713.15	1096.15	---	---	---	---	---	---	2
324	Tm	THULIUM	168.934	1818.15	2219.15	---	---	---	---	---	---	4,5
325	U	URANIUM	238.029	1408.15	4135.00	---	---	---	---	---	---	4,5
326	UF6	URANIUM FLUORIDE	352.019	342.35	328.85	505.80	46.60	250.00	1.4081	0.277	0.318	2,10
327	V	VANADIUM	50.942	2183.15	3665.00	---	---	---	---	---	---	4,5
328	VCl4	VANADIUM TETRACHLORIDE	192.752	247.45	425.00	697.00	60.30	268.00	0.7192	0.279	0.186	1
329	VOCl3	VANADIUM OXYTRICHLORIDE	173.299	193.65	400.00	636.00	49.96	290.00	0.5976	0.274	---	1,21
330	W	TUNGSTEN	183.840	3695.15	5645.00	14756.0	---	33.90	5.4230	---	---	4,5,6
331	WF6	TUNGSTEN FLUORIDE	297.830	272.65	290.45	468.56	46.75	228.32	1.3044	0.274	---	2,21
332	Xe	XENON	131.290	161.36	165.03	289.74	58.40	118.00	1.1126	0.286	0.000	1
333	Yb	YTTERBIUM	173.040	1097.15	1660.00	---	---	---	---	---	---	4,5
334	Yt	YTTRIUM	88.906	1799.15	3055.00	---	---	---	---	---	---	4,5

* A computer program, containing data for all compounds, is available for a nominal fee (Carl L. Yaws, Box 10053, Lamar University, Beaumont, TX 77710, phone/FAX 409-880-8787). The computer program is in ASCII which can be accessed by other software.

NO	FORMULA	NAME	MW g/mol	T_F K	T_B K	T_C K	P_C bar	V_C cm^3/mol	ρ_C g/cm^3	Z_C	ω	SOURCE
335	Zn	ZINC	65.390	692.70	1181.15	3170.00	2904.0	33.00	1.9815	0.364	0.078	1
336	ZnCl2	ZINC CHLORIDE	136.295	638.15	1005.15	---	---	---	---	---	---	3
337	ZnF2	ZINC FLUORIDE	103.387	1145.15	1770.15	---	---	---	---	---	---	3
338	ZnO	ZINC OXIDE	81.389	2248.20	---	---	---	---	---	---	---	1
339	ZnSO4	ZINC SULFATE	161.454	953.00	---	---	---	---	---	---	---	1
340	Zr	ZIRCONIUM	91.224	2128.15	4598.00	8802.00	---	---	---	---	---	4,5,6
341	ZrBr4	ZIRCONIUM BROMIDE	410.840	723.15	630.15	---	---	---	---	---	---	2
342	ZrCl4	ZIRCONIUM CHLORIDE	233.035	710.15	604.15	---	---	---	---	---	---	2
343	ZrI4	ZIRCONIUM IODIDE	598.842	772.15	704.15	---	---	---	---	---	---	2

* A computer program, containing data for all compounds, is available for a nominal fee (Carl L. Yaws, Box 10053, Lamar University, Beaumont, TX 77710, phone/FAX 409-880-8787). The computer program is in ASCII which can be accessed by other software.

NOTE:

1. Sources for the property data are:

1. Daubert, T. E. and R. P. Danner, DATA COMPILATION OF PROPERTIES OF PURE COMPOUNDS, Parts 1, 2, 3 and 4, Supplements 1 and 2, DIPPR Project, AIChE, New York, NY (1985-1992).
2. Ohe, S., COMPUTER AIDED DATA BOOK OF VAPOR PRESSURE, Data Book Publishing Company, Tokyo, Japan (1976).
3. PERRY'S CHEMICAL ENGINEERING HANDBOOK, 6th ed., McGraw-Hill, New York, NY (1984).
4. Nesmeyanov, A. N., VAPOR PRESSURE OF THE CHEMICAL ELEMENTS, Elsevier, New York, NY (1963).
5. CRC HANDBOOK OF CHEMISTRY AND PHYSICS, 66th - 75th eds., CRC Press, Inc., Boca Raton, FL (1985-1994).
6. Simmrock, K. H., R. Janowsky and A. Ohnsorge, CRITICAL DATA OF PURE SUBSTANCES, Vol. II, Parts 1 and 2, Dechema Chemistry Data Series, 6000 Frankfurt/Main, Germany (1986).
7. CONDENSED CHEMICAL DICTIONARY, 10th (G. G. Hawley) and 11th eds. (N. I. Sax and R. J. Lewis, Jr.), Van Nostrand Reinhold Co., New York, NY (1981,1987).
9. LANGE'S HANDBOOK OF CHEMISTRY, 13th and 14th eds., McGraw-Hill, New York, NY (1985, 1992).
10. Reid, R. C., J. M. Prausnitz and B. E. Poling, THE PROPERTIES OF GASES AND LIQUIDS, 3rd ed. (R. C. Reid and T. K. Sherwood), 4th ed., McGraw-Hill, New York, NY (1977, 1987).
11. Rabinovich, V. A., editor, THERMOPHYSICAL PROPERTIES OF GASES AND LIQUIDS, translated from Russian, U. S. Dept. Commerce, Springfield, VA (1970).
12. Yaws, C. L. and others, Solid State Technology, 16, No. 1, 39 (1973).
13. Yaws, C. L. and others, Solid State Technology, 17, No. 1, 47 (1974).
14. Yaws, C. L. and others, Solid State Technology, 17, No. 11, 31 (1974).
15. Yaws, C. L. and others, Solid State Technology, 18, No. 1, 35 (1975).
16. Yaws, C. L. and others, Solid State Technology, 21, No. 1, 43 (1978).
17. Yaws, C. L. and others, Solid State Technology, 24, No. 1, 87 (1981).
18. Yaws, C. L. and others, J. Ch. I. Ch. E., 12, 33 (1981).
19. Yaws, C. L. and others, J. Ch. I. Ch. E., 14, 205 (1983).
20. Yaws, C. L. and others, Ind. Eng. Chem. Process Des. Dev., 23, 48 (1984).
21. Estimated.

2. Very limited experimental data for critical constants and acentric factor are available for inorganic compounds as compared to the more abundant experimental data which are available for organic compounds. Thus, the estimates for these substances should be considered rough approximations in the absence of experimental data.

Appendix E

HEAT CAPACITY FOR INORGANIC COMPOUNDS AND ELEMENTS**

Carl L. Yaws, Mei Han and Sachin D. Sheth
Lamar University, Beaumont, Texas

$$C_P = A + B\,T + C\,T^2 + D\,T^3 + E\,T^4 \quad (C_P - joule/g\text{-}mol\ K,\ T - K)$$

NO	FORMULA	NAME	A	B	C	D	E	TMIN	TMAX	PHASE
1	Ag	SILVER	24.710	1.1400E-03	3.8800E-06	0.0000E+00	0.0000E+00	298	1234	solid
2	AgCl	SILVER CHLORIDE	-18.821	4.5954E-01	-1.0527E-03	1.1270E-06	-4.6103E-10	298	728	solid
3	AgI	SILVER IODIDE	24.351	1.0083E-01	-9.4018E-11	1.7495E-13	-1.2153E-16	298	420	solid
4	Al	ALUMINUM	14.490	5.1700E-02	-7.9280E-05	4.9600E-08	0.0000E+00	298	933	solid
5	AlB3H12	ALUMINUM BOROHYDRIDE	---	---	---	---	---	--	--	gas
6	AlBr3	ALUMINUM BROMIDE*	39.535	2.0117E-01	-3.2271E-04	2.2542E-07	-5.7081E-11	100	1500	gas
7	AlCl3	ALUMINUM CHLORIDE	34.535	2.0117E-01	-3.2271E-04	2.2542E-07	-5.7081E-11	100	1500	gas
8	AlF3	ALUMINUM FLUORIDE	3.104	4.3453E-01	-8.9929E-04	9.6363E-07	-3.9456E-10	298	727	solid
9	AlI3	ALUMINUM IODIDE	70.626	9.4809E-02	-4.3607E-11	7.6722E-14	-5.0239E-17	298	464	solid
10	Al2O3	ALUMINUM OXIDE	-41.081	6.5255E-01	-1.0383E-03	7.5410E-07	-2.0346E-10	298	1273	solid
11	Al2S3O12	ALUMINUM SULFATE	-36.900	1.4427E+00	-1.6840E-03	7.0100E-07	0.0000E+00	55	900	solid
12	Ar	ARGON	20.786	0.0000E+00	0.0000E+00	0.0000E+00	0.0000E+00	100	1500	gas
13	As	ARSENIC	21.882	9.2885E-03	3.5916E-15	-3.5942E-18	1.2807E-21	298	1090	solid
14	AsBr3	ARSENIC TRIBROMIDE*	68.784	5.9075E-02	-6.9453E-05	3.5616E-08	-6.6261E-12	298	2000	gas
15	AsCl3	ARSENIC TRICHLORIDE	63.784	5.9075E-02	-6.9453E-05	3.5616E-08	-6.6261E-12	298	2000	gas
16	AsF3	ARSENIC TRIFLUORIDE	39.648	1.2752E-01	-1.4842E-04	7.6110E-08	-1.4160E-11	298	2000	gas
17	AsF5	ARSENIC PENTAFLUORIDE*	74.312	1.2752E-01	-1.4842E-04	7.6110E-08	-1.4160E-11	298	2000	gas
18	AsH3	ARSINE	31.578	2.2579E-04	1.2295E-04	-1.3416E-07	4.1378E-11	80	1500	gas
19	AsI3	ARSENIC TRIIODIDE*	71.126	9.4809E-02	-4.3607E-11	7.6722E-14	-5.0239E-17	298	419	solid
20	As2O3	ARSENIC TRIOXIDE	35.020	2.0334E-01	-2.6126E-12	4.3024E-15	-2.6100E-18	273	548	solid
21	At	ASTATINE	29.288	-1.0277E-10	3.6271E-13	-5.5948E-16	3.1852E-19	298	575	solid
22	Au	GOLD	24.282	2.7145E-03	3.7682E-06	-2.6406E-09	6.8561E-13	298	1336	solid
23	B	BORON	-12.195	1.1650E-01	-1.4732E-04	8.5905E-08	-1.8326E-11	298	1700	solid
24	BBr3	BORON TRIBROMIDE	38.762	1.4855E-01	-2.0938E-04	1.3632E-07	-3.3271E-11	298	1500	gas
25	BCl3	BORON TRICHLORIDE	24.444	1.9076E-01	-2.6142E-04	1.6467E-07	-3.8875E-11	100	1500	gas
26	BF3	BORON TRIFLUORIDE	22.487	1.1814E-01	-8.7099E-05	2.2344E-08	1.2182E-13	100	1500	gas
27	BH2CO	BORINE CARBONYL*	-2.568	1.7067E-01	-6.8997E-05	1.2108E-08	-7.6892E-13	298	1500	gas
28	BH3O3	BORIC ACID	90.800	0.0000E+00	0.0000E+00	0.0000E+00	0.0000E+00	293	303	solid
29	B2D6	DEUTERODIBORANE*	21.184	1.7067E-01	-6.8997E-05	1.2108E-08	-7.6892E-13	100	6000	gas
30	B2H5Br	DIBORANE HYDROBROMIDE*	31.932	1.7067E-01	-6.8997E-05	1.2108E-08	-7.6892E-13	100	6000	gas
31	B2H6	DIBORANE	19.984	1.7067E-01	-6.8997E-05	1.2108E-08	-7.6892E-13	100	6000	gas
32	B3N3H6	BORINE TRIAMINE	-38.941	6.0750E-01	-5.9547E-04	3.0827E-07	-6.4789E-11	298	1500	gas
33	B4H10	TETRABORANE*	-66.873	5.6949E-01	-3.3162E-04	5.6690E-08	7.8563E-12	298	1500	gas
34	B5H9	PENTABORANE	-48.121	5.6949E-01	-3.3162E-04	5.6690E-08	7.8563E-12	298	1500	gas
35	B5H11	TETRAHYDROPENTABORANE*	-47.121	5.6949E-01	-3.3162E-04	5.6690E-08	7.8563E-12	298	1500	gas
36	B10H14	DECABORANE	---	---	---	---	---	--	--	solid
37	Ba	BARIUM	20.439	2.6930E-02	-3.4202E-05	3.9641E-08	-1.7649E-11	298	643	solid
38	Be	BERYLLIUM	6.512	5.4118E-02	-6.4462E-05	4.0254E-08	-9.2384E-12	298	1556	solid
39	BeB2H8	BERYLLIUM BOROHYDRIDE	---	---	---	---	---	--	--	solid
40	BeBr2	BERYLLIUM BROMIDE	-4.229	4.7593E-01	-1.0617E-03	1.1042E-06	-4.3793E-10	298	761	solid
41	BeCl2	BERYLLIUM CHLORIDE	-2.329	4.2397E-01	-9.7917E-04	1.0995E-06	-4.7325E-10	298	676	solid
42	BeF2	BERYLLIUM FLUORIDE	-113.564	1.2467E+00	-3.5145E-03	4.7410E-06	-2.4784E-09	298	501	solid
43	BeI2	BERYLLIUM IODIDE	1.403	4.5521E-01	-1.0232E-03	1.0716E-06	-4.2815E-10	298	753	solid
44	Bi	BISMUTH	22.933	1.0125E-02	6.4912E-13	-1.0388E-15	6.1496E-19	298	545	solid
45	BiBr3	BISMUTH TRIBROMIDE	---	---	---	---	---	--	--	solid
46	BiCl3	BISMUTH TRICHLORIDE	66.314	7.8865E-02	-1.4984E-04	1.2946E-07	-4.2087E-11	298	1000	solid
47	BrF5	BROMINE PENTAFLUORIDE	27.183	3.9339E-01	-5.9604E-04	4.0800E-07	-1.0308E-10	298	1500	gas
48	Br2	BROMINE	27.169	4.9172E-02	-8.5027E-05	6.2796E-08	-1.6556E-11	100	1500	gas
49	C	CARBON	-7.353	6.9494E-02	-6.4040E-05	2.8510E-08	-4.9877E-12	301	1700	solid
50	CCl2O	PHOSGENE	20.747	1.7972E-01	-2.3242E-04	1.4224E-07	-3.3087E-11	100	1500	gas
51	CF2O	CARBONYL FLUORIDE	23.640	8.9853E-02	-2.4575E-05	-2.8140E-08	1.4023E-11	100	1500	gas
52	CH4N2O	UREA	17.250	2.3180E-01	7.9000E-05	0.0000E+00	0.0000E+00	80	400	solid
53	CH4N2S	THIOUREA	21.530	2.2204E-01	-1.7193E-04	7.4203E-08	-1.3867E-11	273	1500	gas
54	CNBr	CYANOGEN BROMIDE	31.562	7.7072E-02	-1.0251E-04	7.0456E-08	-1.8400E-11	298	1500	gas
55	CNCl	CYANOGEN CHLORIDE	21.270	1.1915E-01	-1.6822E-04	1.1457E-07	-2.9210E-11	100	1500	gas
56	CNF	CYANOGEN FLUORIDE	26.132	7.5002E-02	-8.3145E-05	4.9592E-08	-1.1885E-11	298	1500	gas
57	CO	CARBON MONOXIDE	29.556	-6.5807E-03	2.0130E-05	-1.2227E-08	2.2617E-12	60	1500	gas
58	COS	CARBONYL SULFIDE	20.913	9.2794E-02	-9.7014E-05	5.0943E-08	-1.0615E-11	100	1500	gas

** A computer program, containing data for all compounds, is available for a nominal fee (Carl L. Yaws, Box 10053, Lamar University, Beaumont, TX 77710, phone/FAX 409-880-8787). The computer program is in ASCII which can be accessed by other software.

$$C_P = A + B T + C T^2 + D T^3 + E T^4 \quad (C_P - joule/g\text{-}mol\ K,\ T - K)$$

NO	FORMULA	NAME	A	B	C	D	E	TMIN	TMAX	PHASE
59	COSe	CARBON OXYSELENIDE*	21.912	9.2794E-02	-9.7014E-05	5.0943E-08	-1.0615E-11	100	1500	gas
60	CO2	CARBON DIOXIDE	27.437	4.2315E-02	-1.9555E-05	3.9968E-09	-2.9872E-13	50	5000	gas
61	CS2	CARBON DISULFIDE	20.461	1.2299E-01	-1.6184E-04	1.0199E-07	-2.4444E-11	100	1500	gas
62	CSeS	CARBON SELENOSULFIDE*	21.461	1.2299E-01	-1.6184E-04	1.0199E-07	-2.4444E-11	100	1500	gas
63	C2N2	CYANOGEN	32.265	1.1687E-01	-1.4171E-04	9.2703E-08	-2.3760E-11	298	1500	gas
64	C3S2	CARBON SUBSULFIDE	---	---	---	---	---	--	--	gas
65	Ca	CALCIUM	21.924	1.4644E-02	1.1410E-14	-1.4961E-17	7.1522E-21	298	737	solid
66	CaF2	CALCIUM FLUORIDE	53.980	5.8600E-02	-3.8180E-05	1.5630E-08	0.0000E+00	298	1400	solid
67	CbF5	COLUMBIUM FLUORIDE	---	---	---	---	---	--	--	solid
68	Cd	CADMIUM	22.217	1.2301E-02	-8.8357E-13	1.3385E-15	-7.4720E-19	298	594	solid
69	CdCl2	CADMIUM CHLORIDE	66.944	3.2217E-02	-1.2143E-14	1.4595E-17	-6.3455E-21	298	841	solid
70	CdF2	CADMIUM FLUORIDE	---	---	---	---	---	--	--	solid
71	CdI2	CADMIUM IODIDE	---	---	---	---	---	--	--	solid
72	CdO	CADMIUM OXIDE	40.376	8.7027E-03	-4.5035E-16	4.2302E-19	-1.4057E-22	298	1200	solid
73	ClF	CHLORINE MONOFLUORIDE	22.567	4.7581E-02	-6.3572E-05	3.9963E-08	-9.4968E-12	298	1500	gas
74	ClFO3	PERCHLORYL FLUORIDE	13.200	2.3797E-01	-2.5150E-04	1.2324E-07	-2.2897E-11	100	1500	gas
75	ClF3	CHLORINE TRIFLUORIDE	21.386	2.2286E-01	-3.3105E-04	2.2357E-07	-5.5964E-11	298	1500	gas
76	ClF5	CHLORINE PENTAFLUORIDE	15.530	4.3077E-01	-6.4695E-04	4.4026E-07	-1.1080E-10	298	1500	gas
77	ClHO3S	CHLOROSULFONIC ACID	21.765	2.7543E-01	-3.3639E-04	2.0259E-07	-4.6684E-11	300	1500	gas
78	ClHO4	PERCHLORIC ACID	73.220	0.0000E+00	0.0000E+00	0.0000E+00	0.0000E+00	298	303	gas
79	ClO2	CHLORINE DIOXIDE	30.482	3.9797E-02	4.5262E-06	-3.2447E-08	1.3089E-11	50	1500	gas
80	Cl2	CHLORINE	27.213	3.0426E-02	-3.3353E-05	1.5961E-08	-2.7021E-12	50	1500	gas
81	Cl2O	CHLORINE MONOXIDE	25.608	1.1593E-01	-1.7038E-04	1.1417E-07	-2.8420E-11	298	1500	gas
82	Cl2O7	CHLORINE HEPTOXIDE*	110.489	3.9797E-02	4.5262E-06	-3.2447E-08	1.3089E-11	50	1500	gas
83	Co	COBALT	19.832	1.6736E-02	3.6244E-14	-4.9839E-17	2.5032E-20	298	700	solid
84	CoCl2	COBALT CHLORIDE	60.291	6.1086E-02	1.8571E-16	-1.9975E-19	7.7177E-23	298	1013	solid
85	CoNC3O4	COBALT NITROSYL TRICARBONYL*	76.352	8.9853E-02	-2.4575E-05	-2.8140E-08	1.4023E-11	100	1500	gas
86	Cr	CHROMIUM	19.044	1.5064E-02	-2.4696E-06	1.1830E-09	-2.0476E-13	298	2176	solid
87	CrC6O6	CHROMIUM CARBONYL	251.667	1.1009E-05	-5.0497E-08	1.0276E-10	-7.8280E-14	293	363	solid
88	CrO2Cl2	CHROMIUM OXYCHLORIDE*	61.375	8.9853E-02	-2.4575E-05	-2.8140E-08	1.4023E-11	100	1500	gas
89	Cs	CESIUM	32.163	0.0000E+00	0.0000E+00	0.0000E+00	0.0000E+00	298	300	solid
90	CsBr	CESIUM BROMIDE	48.530	1.0837E-02	-7.4433E-15	8.4228E-18	-3.4349E-21	298	908	solid
91	CsCl	CESIUM CHLORIDE	45.857	2.2096E-02	-1.3165E-13	1.7225E-16	-8.2107E-20	298	743	solid
92	CsF	CESIUM FLUORIDE	46.685	1.7744E-02	1.8485E-15	-2.0046E-18	7.7924E-22	298	976	solid
93	CsI	CESIUM IODIDE	77.717	-1.9833E-01	4.9279E-04	-4.6011E-07	1.6255E-10	298	894	solid
94	Cu	COPPER	22.635	6.2760E-03	1.5835E-16	-1.3498E-19	4.0395E-23	298	1357	solid
95	CuBr	CUPROUS BROMIDE	49.831	1.6610E-02	-6.5590E-14	9.4279E-17	-4.9688E-20	298	653	solid
96	CuCl	CUPROUS CHLORIDE	38.277	3.4969E-02	0.0000E+00	0.0000E+00	0.0000E+00	298	703	solid
97	CuCl2	CUPRIC CHLORIDE	63.320	3.4880E-02	-1.6050E-05	0.0000E+00	0.0000E+00	300	906	solid
98	CuI	COPPER IODIDE	50.626	1.1966E-02	-1.2465E-14	1.4700E-17	-6.2634E-21	298	861	solid
99	DCN	DEUTERIUM CYANIDE*	25.967	3.7969E-02	-1.2416E-05	-3.2240E-09	2.2610E-12	100	1500	gas
100	D2	DEUTERIUM	31.159	-1.2796E-02	2.4964E-05	-1.5015E-08	3.3248E-12	100	1500	gas
101	D2O	DEUTERIUM OXIDE	33.308	-4.6722E-03	3.4878E-05	-2.2602E-08	4.4864E-12	100	2000	gas
102	Eu	EUROPIUM	24.309	8.2843E-03	3.9790E-15	-3.9528E-18	1.3976E-21	298	1100	solid
103	F2	FLUORINE	27.408	1.2928E-02	7.0701E-06	-1.6302E-08	5.9789E-12	100	1500	gas
104	F2O	FLUORINE OXIDE	16.655	1.3539E-01	-1.8807E-04	1.2034E-07	-2.8760E-11	298	1500	gas
105	Fe	IRON	25.913	-2.9321E-02	1.1358E-04	-9.5921E-08	3.0427E-11	298	1033	solid
106	FeC5O5	IRON PENTACARBONYL*	86.030	8.9853E-02	-2.4575E-05	-2.8140E-08	1.4023E-11	100	1500	gas
107	FeCl2	FERROUS CHLORIDE	51.564	1.4293E-01	-2.6346E-04	2.3592E-07	-7.9693E-11	298	950	solid
108	FeCl3	FERRIC CHLORIDE	62.342	1.1506E-01	1.7739E-12	-2.7316E-15	1.5522E-18	298	577	solid
109	Fr	FRANCIUM	34.051	-4.3909E-02	2.8935E-04	-7.9906E-07	7.9673E-10	298	300	solid
110	Ga	GALLIUM	27.824	-3.2136E-14	4.2017E-17	-2.1331E-20	3.6846E-24	303	2520	liquid
111	GaCl3	GALLIUM TRICHLORIDE	118.410	0.0000E+00	0.0000E+00	0.0000E+00	0.0000E+00	298	351	solid
112	Gd	GADOLINIUM	38.581	-5.0836E-03	-3.7104E-16	2.7645E-19	-7.1448E-23	298	1623	solid
113	Ge	GERMANIUM	14.289	4.8727E-02	-7.5912E-05	5.7125E-08	-1.6010E-11	298	1213	solid
114	GeBr4	GERMANIUM BROMIDE	87.753	7.3780E-02	-1.0615E-04	6.8167E-08	-1.6121E-11	298	1500	gas
115	GeCl4	GERMANIUM CHLORIDE	69.397	1.3977E-01	-2.0098E-04	1.2907E-07	-3.0523E-11	298	1500	gas
116	GeHCl3	TRICHLORO GERMANE*	24.939	2.5068E-01	-3.4090E-04	2.1707E-07	-5.2003E-11	100	1500	gas
117	GeH4	GERMANE	-15.224	3.0554E-01	-4.0678E-04	2.6063E-07	-6.1884E-11	200	1500	gas
118	Ge2H6	DIGERMANE*	21.867	3.0554E-01	-4.0678E-04	2.6063E-07	-6.1884E-11	200	1500	gas
119	Ge3H8	TRIGERMANE*	63.478	3.0554E-01	-4.0678E-04	2.6063E-07	-6.1884E-11	200	1500	gas
120	HBr	HYDROGEN BROMIDE	30.169	-8.0274E-03	1.6731E-05	-7.4730E-09	8.3068E-13	200	1500	gas
121	HCN	HYDROGEN CYANIDE	25.766	3.7969E-02	-1.2416E-05	-3.2240E-09	2.2610E-12	100	1500	gas
122	HCl	HYDROGEN CHLORIDE	29.244	-1.2615E-03	1.1210E-06	4.9676E-09	-2.4963E-12	50	1500	gas
123	HF	HYDROGEN FLUORIDE	29.085	9.6118E-04	-4.4705E-06	6.7830E-09	-2.1975E-12	50	1500	gas
124	HI	HYDROGEN IODIDE	29.770	-7.4945E-03	2.0687E-05	-1.1963E-08	2.1010E-12	100	1500	gas
125	HNO3	NITRIC ACID	19.755	1.3415E-01	-6.1116E-05	-1.2343E-08	1.1106E-11	100	1500	gas

** A computer program, containing data for all compounds, is available for a nominal fee (Carl L. Yaws, Box 10053, Lamar University, Beaumont, TX 77710, phone/FAX 409-880-8787). The computer program is in ASCII which can be accessed by other software.

$$C_P = A + B\,T + C\,T^2 + D\,T^3 + E\,T^4 \quad (C_P - \text{joule/g-mol K}, \ T - K)$$

NO	FORMULA	NAME	A	B	C	D	E	TMIN	TMAX	PHASE
126	H2	HYDROGEN	25.399	2.0178E-02	-3.8549E-05	3.1880E-08	-8.7585E-12	250	1500	gas
127	H2O	WATER	33.933	-8.4186E-03	2.9906E-05	-1.7825E-08	3.6934E-12	100	1500	gas
128	H2O2	HYDROGEN PEROXIDE	36.181	8.2657E-03	6.6420E-05	-6.9944E-08	2.0951E-11	100	1500	gas
129	H2S	HYDROGEN SULFIDE	33.878	-1.1216E-02	5.2578E-05	-3.8397E-08	9.0281E-12	100	1500	gas
130	H2SO4	SULFURIC ACID	9.486	3.3795E-01	-3.8078E-04	2.1308E-07	-4.6878E-11	100	1500	gas
131	H2S2	HYDROGEN DISULFIDE*	58.617	-1.1216E-02	5.2578E-05	-3.8397E-08	9.0281E-12	100	1500	gas
132	H2Se	HYDROGEN SELENIDE*	34.878	-1.1216E-02	5.2578E-05	-3.8397E-08	9.0281E-12	100	1500	gas
133	H2Te	HYDROGEN TELLURIDE*	34.878	-1.1216E-02	5.2578E-05	-3.8397E-08	9.0281E-12	100	1500	gas
134	H3NO3S	SULFAMIC ACID	129.000	0.0000E+00	0.0000E+00	0.0000E+00	0.0000E+00	293	303	solid
135	He	HELIUM-3	20.786	0.0000E+00	0.0000E+00	0.0000E+00	0.0000E+00	100	1500	gas
136	He	HELIUM-4	20.786	0.0000E+00	0.0000E+00	0.0000E+00	0.0000E+00	100	1500	gas
137	Hf	HAFNIUM	25.703	0.0000E+00	0.0000E+00	0.0000E+00	0.0000E+00	298	308	solid
138	Hg	MERCURY	31.106	-1.4350E-02	1.2908E-05	0.0000E+00	0.0000E+00	234	630	liquid
139	HgBr2	MERCURIC BROMIDE	66.584	2.9288E-02	6.0558E-12	-1.0022E-14	6.1505E-18	298	514	solid
140	HgCl2	MERCURIC CHLORIDE	52.426	1.3494E-01	-3.1212E-04	3.9829E-07	-1.9638E-10	298	550	solid
141	HgI2	MERCURIC IODIDE	56.286	3.4249E-01	-8.7036E-04	1.0951E-06	-5.3193E-10	298	563	solid
142	IF7	IODINE HEPTAFLUORIDE	38.537	5.0269E-01	-7.1344E-04	4.5817E-07	-1.0835E-10	298	1500	gas
143	I2	IODINE	34.150	1.3931E-02	-2.0953E-05	1.4362E-08	-3.5950E-12	100	1500	gas
144	In	INDIUM	21.506	1.7573E-02	-2.4686E-11	4.5400E-14	-3.1151E-17	298	429	solid
145	Ir	IRIDIUM	23.347	5.7739E-03	-3.4328E-17	2.3440E-20	-5.5160E-24	298	1809	solid
146	K	POTASSIUM	7.837	7.1983E-02	---	---	---	298	336	solid
147	KBr	POTASSIUM BROMIDE	34.390	1.1766E-01	-2.9798E-04	2.6546E-07	-8.5845E-11	298	1007	solid
148	KCl	POTASSIUM CHLORIDE	39.680	5.8380E-02	-8.1210E-05	5.0100E-08	0.0000E+00	200	1000	solid
149	KF	POTASSIUM FLUORIDE	45.982	1.4422E-02	3.8443E-15	-3.7237E-18	1.2824E-21	298	1130	solid
150	KI	POTASSIUM IODIDE	66.603	-1.0541E-01	2.6297E-04	-2.3479E-07	7.9064E-11	298	954	solid
151	KOH	POTASSIUM HYDROXIDE	-4.100	4.6870E-01	-1.0706E-03	9.4200E-07	0.0000E+00	50	522	solid
152	Kr	KRYPTON	20.786	0.0000E+00	0.0000E+00	0.0000E+00	0.0000E+00	100	6200	gas
153	La	LANTHANUM	25.815	6.9944E-03	-2.6336E-16	2.4699E-19	-8.2601E-23	298	1141	solid
154	Li	LITHIUM	-8.997	3.3378E-01	-1.3341E-03	2.4813E-06	-1.6521E-09	100	454	solid
155	LiBr	LITHIUM BROMIDE	68.390	-1.5919E-01	4.3042E-04	-4.2500E-07	1.5943E-10	298	823	solid
156	LiCl	LITHIUM CHLORIDE	41.417	2.3397E-02	-1.3477E-14	1.5769E-17	-6.6511E-21	298	883	solid
157	LiF	LITHIUM FLUORIDE	16.725	1.3208E-01	-2.0239E-04	1.6126E-07	-4.8053E-11	298	1121	solid
158	LiI	LITHIUM IODIDE	41.903	2.8091E-02	6.6308E-14	-8.7334E-17	4.1888E-20	298	742	solid
159	Lu	LUTECIUM	25.104	6.2760E-03	-6.3338E-17	4.0169E-20	-8.7163E-24	298	2000	solid
160	Mg	MAGNESIUM	19.801	2.2569E-02	-2.4620E-05	2.2489E-08	-7.7607E-12	298	923	solid
161	MgCl2	MAGNESIUM CHLORIDE	31.946	2.2965E-01	-4.2858E-04	3.7372E-07	-1.2269E-10	298	987	solid
162	MgO	MAGNESIUM OXIDE	-8.000	2.4690E-01	-3.7660E-04	1.9830E-07	0.0000E+00	200	800	solid
163	Mn	MANGANESE	15.292	5.4688E-02	-7.7529E-05	6.7461E-08	-2.2097E-11	298	990	solid
164	MnCl2	MANGANESE CHLORIDE	42.233	1.7706E-01	-3.2746E-04	2.9913E-07	-1.0323E-10	298	923	solid
165	Mo	MOLYBDENUM	21.715	6.9371E-03	6.4610E-17	-2.9498E-20	4.5417E-24	298	2890	solid
166	MoF6	MOLYBDENUM FLUORIDE	41.680	4.1131E-01	-6.0261E-04	4.0311E-07	-1.0025E-10	298	1500	gas
167	MoO3	MOLYBDENUM OXIDE	30.322	2.3631E-01	-3.7096E-04	3.0598E-07	-9.4632E-11	298	1068	solid
168	NCl3	NITROGEN TRICHLORIDE	30.253	1.7432E-01	-2.1734E-04	1.1642E-07	-2.2436E-11	100	2000	gas
169	ND3	HEAVY AMMONIA*	34.574	-1.2581E-02	8.8906E-05	-7.1783E-08	1.8569E-11	100	1500	gas
170	NF3	NITROGEN TRIFLUORIDE	18.732	1.5505E-01	-1.4305E-04	5.3741E-08	-5.8443E-12	100	1500	gas
171	NH3	AMMONIA	33.573	-1.2581E-02	8.8906E-05	-7.1783E-08	1.8569E-11	100	1500	gas
172	NH3O	HYDROXYLAMINE	21.935	1.0340E-01	-5.8693E-05	1.0557E-08	1.5150E-12	200	1500	gas
173	NH4Br	AMMONIUM BROMIDE	95.159	-2.5788E-03	1.3674E-05	-3.2219E-08	2.8462E-11	273	293	solid
174	NH4Cl	AMMONIUM CHLORIDE	34.757	1.1150E-01	0.0000E+00	0.0000E+00	0.0000E+00	458	700	solid
175	NH4I	AMMONIUM IODIDE	60.291	7.1764E-02	-7.8705E-14	9.6400E-17	-4.2746E-20	298	824	solid
176	NH5O	AMMONIUM HYDROXIDE	---	---	---	---	---	--	--	liquid
177	NH5S	AMMONIUM HYDROGENSULFIDE	---	---	---	---	---	--	--	solid
178	NO	NITRIC OXIDE	33.227	-2.3626E-02	5.3156E-05	-3.7858E-08	9.1197E-12	50	1500	gas
179	NOCl	NITROSYL CHLORIDE	28.551	7.5899E-02	-9.4410E-05	6.0476E-08	-1.5054E-11	100	1500	gas
180	NOF	NITROSYL FLUORIDE*	27.551	7.5899E-02	-9.4410E-05	6.0476E-08	-1.5054E-11	100	1500	gas
181	NO2	NITROGEN DIOXIDE	32.791	-7.4294E-04	8.1722E-05	-8.2872E-08	2.4424E-11	50	1500	gas
182	N2	NITROGEN	29.414	-4.5993E-03	1.3004E-05	-5.4759E-09	2.9239E-13	50	1500	gas
183	N2F4	TETRAFLUOROHYDRAZINE	12.422	3.0609E-01	-3.1077E-04	1.3914E-07	-2.2235E-11	100	1500	gas
184	N2H4	HYDRAZINE	23.630	9.1270E-02	2.9042E-05	-7.1858E-08	2.5093E-11	100	1500	gas
185	N2H4C	AMMONIUM CYANIDE*	52.812	9.1270E-02	2.9042E-05	-7.1858E-08	2.5093E-11	100	1500	gas
186	N2H6CO2	AMMONIUM CARBAMATE	---	---	---	---	---	--	--	solid
187	N2O	NITROUS OXIDE	23.219	6.1984E-02	-3.7989E-05	6.9671E-09	8.1421E-13	100	1500	gas
188	N2O3	NITROGEN TRIOXIDE	28.509	1.6895E-01	-1.8161E-04	9.9662E-08	-2.1975E-11	100	1500	gas
189	N2O4	NITROGEN TETRAOXIDE	29.587	2.2719E-01	-2.2740E-04	1.0698E-07	-1.9223E-11	50	1500	gas
190	N2O5	NITROGEN PENTOXIDE	63.710	1.2317E-01	-5.9937E-05	1.1842E-08	-8.1522E-13	200	6000	gas
191	Na	SODIUM	14.791	4.4228E-02	4.6243E-09	-9.2264E-12	6.8903E-15	298	371	solid
192	NaBr	SODIUM BROMIDE	41.654	4.4400E-02	-4.6524E-05	2.1650E-08	0.0000E+00	200	1000	solid

$$C_P = A + B T + C T^2 + D T^3 + E T^4 \quad (C_P - \text{joule/g-mol K}, \ T - K)$$

NO	FORMULA	NAME	A	B	C	D	E	TMIN	TMAX	PHASE
193	NaCN	SODIUM CYANIDE	68.399	7.7953E-04	0.0000E+00	0.0000E+00	0.0000E+00	298	835	solid
194	NaCl	SODIUM CHLORIDE	36.710	6.2770E-02	-6.6670E-05	2.8000E-08	0.0000E+00	200	1074	solid
195	NaF	SODIUM FLUORIDE	18.360	1.6682E-01	-3.1756E-04	2.7570E-07	-8.3700E-11	200	1200	solid
196	NaI	SODIUM IODIDE	48.877	1.2054E-02	-3.1542E-15	3.5205E-18	-1.4126E-21	298	933	solid
197	NaOH	SODIUM HYDROXIDE	-31.800	8.4550E-01	-3.0665E-03	5.0706E-06	-2.9200E-09	60	592	solid
198	Na2SO4	SODIUM SULFATE	-8.040	9.3400E-01	-2.2370E-03	2.1350E-06	0.0000E+00	59	458	solid
199	Nb	NIOBIUM	23.723	4.0166E-03	-2.2179E-17	1.0763E-20	-1.7590E-24	298	2740	solid
200	Nd	NEODYMIUM	36.372	-6.8306E-02	1.6672E-04	-1.3166E-07	3.8858E-11	298	1135	solid
201	Ne	NEON	20.786	0.0000E+00	0.0000E+00	0.0000E+00	0.0000E+00	100	1500	gas
202	Ni	NICKEL	-13.157	2.7626E-01	-7.0013E-04	8.2194E-07	-3.7096E-10	298	630	solid
203	NiC4O4	NICKEL CARBONYL*	69.830	8.9853E-02	-2.4575E-05	-2.8140E-08	1.4023E-11	100	1500	gas
204	NiF2	NICKEL FLUORIDE	72.258	1.0293E-02	5.2390E-17	-2.7624E-20	4.9116E-24	298	1723	solid
205	Np	NEPTUNIUM	---	---	---	---	---	--	--	solid
206	O2	OXYGEN	29.526	-8.8999E-03	3.8083E-05	-3.2629E-08	8.8607E-12	50	1500	gas
207	O3	OZONE	31.467	1.4982E-02	6.7966E-05	-8.4157E-08	2.7205E-11	50	1500	gas
208	Os	OSMIUM	23.556	3.8493E-03	4.0482E-17	-2.6647E-20	6.0441E-24	298	1877	solid
209	OsOF5	OSMIUM OXIDE PENTAFLUORIDE	---	---	---	---	---	--	--	gas
210	OsO4	OSMIUM TETROXIDE - YELLOW	24.585	2.4934E-01	-3.3236E-04	2.1344E-07	-5.0476E-11	298	1500	gas
211	OsO4	OSMIUM TETROXIDE - WHITE	24.585	2.4934E-01	-3.3236E-04	2.1344E-07	-5.0476E-11	298	1500	gas
212	P	PHOSPHORUS - WHITE	4.400	1.2200E-01	-2.5300E-04	2.2765E-07	-7.1400E-11	31	317	solid
213	PBr3	PHOSPHORUS TRIBROMIDE	56.758	1.0427E-01	-1.6430E-04	1.1542E-07	-2.9670E-11	298	1500	gas
214	PCl2F3	PHOSPHORUS DICHLORIDE TRIFLUORIDE*	20.696	4.7099E-01	-7.8406E-04	5.6105E-07	-1.4442E-10	100	1500	gas
215	PCl3	PHOSPHORUS TRICHLORIDE	27.213	2.4066E-01	-3.9532E-04	2.8032E-07	-7.1695E-11	100	1500	gas
216	PCl5	PHOSPHORUS PENTACHLORIDE	25.701	4.7099E-01	-7.8406E-04	5.6105E-07	-1.4442E-10	100	1500	gas
217	PH3	PHOSPHINE	32.964	-1.4201E-02	1.3216E-04	-1.1915E-07	3.2843E-11	100	1500	gas
218	PH4Br	PHOSPHONIUM BROMIDE*	62.034	-1.4201E-02	1.3216E-04	-1.1915E-07	3.2843E-11	100	1500	gas
219	PH4Cl	PHOSPHONIUM CHLORIDE*	61.942	-1.4201E-02	1.3216E-04	-1.1915E-07	3.2843E-11	100	1500	gas
220	PH4I	PHOSPHONIUM IODIDE*	62.037	-1.4201E-02	1.3216E-04	-1.1915E-07	3.2843E-11	100	1500	gas
221	POCl3	PHOSPHORUS OXYCHLORIDE	23.911	3.2446E-01	-5.0571E-04	3.4836E-07	-8.7607E-11	40	1500	gas
222	PSBr3	PHOSPHORUS THIOBROMIDE	63.322	1.6822E-01	-2.5688E-04	1.7681E-07	-4.4843E-11	298	1500	gas
223	PSCl3	PHOSPHORUS THIOCHLORIDE	27.454	3.3554E-01	-5.4132E-04	3.7986E-07	-9.6538E-11	100	1500	gas
224	P4O6	PHOSPHORUS TRIOXIDE	-26.248	8.7464E-01	-1.2429E-03	8.1470E-07	-1.9988E-10	298	1500	gas
225	P4O10	PHOSPHORUS PENTOXIDE	9.600	5.9230E-01	1.3880E-03	-3.6600E-06	0.0000E+00	20	325	solid
226	P4S10	PHOSPHORUS PENTASULFIDE	-25.525	2.3950E+00	-6.9192E-03	9.9713E-06	-5.3678E-09	30	560	solid
227	Pb	LEAD	23.552	9.7404E-03	-2.9945E-14	4.4267E-17	-2.4109E-20	298	601	solid
228	PbBr2	LEAD BROMIDE	77.781	9.2048E-03	2.1995E-13	-3.1856E-16	1.6938E-19	298	640	solid
229	PbCl2	LEAD CHLORIDE	60.760	4.1535E-02	1.4883E-14	-1.9009E-17	8.8300E-21	298	768	solid
230	PbF2	LEAD FLUORIDE	54.940	6.4425E-02	-9.8385E-13	1.4615E-15	-7.9884E-19	298	613	solid
231	PbI2	LEAD IODIDE	75.312	1.9665E-02	-3.7105E-13	5.1367E-16	-2.6034E-19	298	680	solid
232	PbO	LEAD OXIDE	41.455	1.5330E-02	6.9602E-14	-8.9876E-17	4.2208E-20	298	762	solid
233	PbS	LEAD SULFIDE	46.434	1.0263E-02	-3.9829E-16	3.2805E-19	-9.5043E-23	298	1387	solid
234	Pd	PALLADIUM	24.225	5.7488E-03	-7.4914E-17	5.1613E-20	-1.2204E-23	298	1823	solid
235	Po	POLONIUM	20.292	2.0418E-02	2.7009E-12	-4.4052E-15	2.6616E-18	298	527	solid
236	Pt	PLATINUM	24.250	5.3764E-03	-9.8794E-17	6.0617E-20	-1.2748E-23	298	2043	solid
237	Ra	RADIUM	20.920	2.0920E-02	-3.7799E-15	4.0654E-18	-1.5696E-21	298	973	solid
238	Rb	RUBIDIUM	13.736	5.6681E-02	3.5243E-06	-7.7226E-09	6.3452E-12	298	312	solid
239	RbBr	RUBIDIUM BROMIDE	48.534	1.0669E-02	-5.0051E-15	5.6215E-18	-2.2585E-21	283	953	solid
240	RbCl	RUBIDIUM CHLORIDE	48.116	1.0418E-02	-4.8316E-15	5.2698E-18	-2.0526E-21	283	988	solid
241	RbF	RUBIDIUM FLUORIDE	59.623	-8.2079E-02	2.2230E-04	-1.8584E-07	5.8308E-11	298	1048	solid
242	RbI	RUBIDIUM IODIDE	48.534	1.1004E-02	-8.4558E-16	1.0369E-18	-4.5189E-22	283	913	solid
243	Re	RHENIUM	24.695	2.7522E-03	2.2393E-06	-7.2768E-10	8.3708E-14	298	3453	solid
244	Re2O7	RHENIUM HEPTOXIDE	113.266	6.8782E-01	-3.3492E-03	7.2466E-06	-5.8789E-09	298	318	solid
245	Rh	RHODIUM	21.966	1.0042E-02	2.6411E-17	-1.5094E-20	2.9416E-24	298	2239	solid
246	Rn	RADON	20.794	-3.9576E-15	4.4438E-18	-1.9112E-21	2.7806E-25	298	3000	gas
247	Ru	RUTHENIUM	21.966	6.2760E-03	5.4587E-16	-4.7849E-19	1.4756E-22	298	1308	solid
248	RuF5	RUTHENIUM PENTAFLUORIDE	---	---	---	---	---	--	--	solid
249	S	SULFUR	25.639	-7.9870E-03	4.7860E-06	-9.5700E-10	0.0000E+00	273	1500	gas
250	SF4	SULFUR TETRAFLUORIDE	15.486	3.1315E-01	-4.4453E-04	2.9083E-07	-7.1213E-11	298	1500	gas
251	SF6	SULFUR HEXAFLUORIDE	-7.934	5.1224E-01	-6.4878E-04	3.7509E-07	-8.1524E-11	100	1500	gas
252	SOBr2	THIONYL BROMIDE	48.491	1.0815E-01	-1.4791E-04	9.4987E-08	-2.2463E-11	298	1500	gas
253	SOCl2	THIONYL CHLORIDE	34.838	1.4750E-01	-1.7837E-04	9.4399E-08	-1.8111E-11	100	2000	gas
254	SOF2	SULFUROUS OXYFLUORIDE	18.639	1.9505E-01	-2.5689E-04	1.5983E-07	-3.7838E-11	298	1500	gas
255	SO2	SULFUR DIOXIDE	29.637	3.4735E-02	9.2903E-06	-2.9885E-08	1.0937E-11	100	1500	gas
256	SO2Cl2	SULFURYL CHLORIDE	18.553	2.9713E-01	-4.2391E-04	2.7784E-07	-6.7857E-11	100	1500	gas
257	SO3	SULFUR TRIOXIDE	22.466	1.1981E-01	-9.0842E-05	2.5503E-08	-7.9208E-13	100	1500	gas
258	S2Cl2	SULFUR MONOCHLORIDE	51.240	1.1549E-01	-1.6270E-04	1.0449E-07	-2.4709E-11	298	1500	gas
259	Sb	ANTIMONY	22.343	8.9538E-03	-8.4656E-15	9.6707E-18	-3.9802E-21	298	903	solid

** A computer program, containing data for all compounds, is available for a nominal fee (Carl L. Yaws, Box 10053, Lamar University, Beaumont, TX 77710, phone/FAX 409-880-8787). The computer program is in ASCII which can be accessed by other software.

$$C_P = A + B T + C T^2 + D T^3 + E T^4 \quad (C_P \text{ - joule/g-mol K}, \ T \text{ - K})$$

NO	FORMULA	NAME	A	B	C	D	E	TMIN	TMAX	PHASE
260	SbBr3	ANTIMONY TRIBROMIDE	71.965	1.2259E-01	1.6687E-09	-3.4808E-12	2.7134E-15	273	368	solid
261	SbCl3	ANTIMONY TRICHLORIDE	63.682	7.1912E-02	-1.0441E-04	6.7050E-08	-1.5856E-11	298	1500	gas
262	SbCl5	ANTIMONY PENTACHLORIDE*	106.953	7.1912E-02	-1.0441E-04	6.7050E-08	-1.5856E-11	298	1500	gas
263	SbH3	STIBINE	13.058	1.2924E-01	-1.3034E-04	6.6840E-08	-1.2435E-11	298	2000	gas
264	SbI3	ANTIMONY TRIIODIDE	71.128	8.8701E-02	-1.4010E-10	2.5303E-13	-1.7032E-16	298	444	solid
265	Sb2O3	ANTIMONY TRIOXIDE	79.914	7.1546E-02	-1.8105E-14	2.1989E-17	-9.6188E-21	273	846	solid
266	Sc	SCANDIUM	23.765	4.5187E-03	4.9739E-17	-3.5288E-20	8.7092E-24	298	1673	solid
267	Se	SELENIUM	15.987	3.0200E-02	-1.7122E-11	3.2998E-14	-2.3673E-17	273	423	solid
268	SeCl4	SELENIUM TETRACHLORIDE*	35.672	3.1545E-01	-5.0601E-04	3.5370E-07	-8.9635E-11	100	1500	gas
269	SeF6	SELENIUM HEXAFLUORIDE*	-6.934	5.1224E-01	-6.4878E-04	3.7509E-07	-8.1524E-11	100	1500	gas
270	SeOCl2	SELENIUM OXYCHLORIDE*	35.838	1.4750E-01	-1.7837E-04	9.4399E-08	-1.8111E-11	100	2000	gas
271	SeO2	SELENIUM DIOXIDE	---	---	---	---	---	--	--	solid
272	Si	SILICON	10.435	4.6948E-02	-5.7687E-05	3.3872E-08	-7.2792E-12	298	1685	solid
273	SiBrCl2F	BROMODICHLOROFLUOROSILANE*	46.876	2.1385E-01	-3.2354E-04	2.2226E-07	-5.6419E-11	298	1500	gas
274	SiBrF3	TRIFLUOROBROMOSILANE*	26.673	2.6781E-01	-3.7339E-04	2.4125E-07	-5.8565E-11	298	1500	gas
275	SiBr2ClF	DIBROMOCHLOROFLUOROSILANE*	57.826	2.1385E-01	-3.2354E-04	2.2226E-07	-5.6419E-11	298	1500	gas
276	SiClF3	TRIFLUOROCHLOROSILANE	26.816	2.6781E-01	-3.7339E-04	2.4125E-07	-5.8565E-11	298	1500	gas
277	SiCl2F2	DICHLORODIFLUOROSILANE*	40.896	2.1385E-01	-3.2354E-04	2.2226E-07	-5.6419E-11	298	1500	gas
278	SiCl3F	TRICHLOROFLUOROSILANE	49.698	2.1385E-01	-3.2354E-04	2.2226E-07	-5.6419E-11	298	1500	gas
279	SiCl4	SILICON TETRACHLORIDE	31.672	3.1545E-01	-5.0601E-04	3.5370E-07	-8.9635E-11	100	1500	gas
280	SiF4	SILICON TETRAFLUORIDE	18.032	2.7359E-01	-3.5566E-04	2.1540E-07	-4.9404E-11	100	1500	gas
281	SiHBr3	TRIBROMOSILANE	29.302	2.5068E-01	-3.4090E-04	2.1707E-07	-5.2003E-11	100	1500	gas
282	SiHCl3	TRICHLOROSILANE	24.939	2.5068E-01	-3.4090E-04	2.1707E-07	-5.2003E-11	100	1500	gas
283	SiHF3	TRIFLUOROSILANE	13.820	2.3386E-01	-2.7143E-04	1.5566E-07	-3.4933E-11	298	1500	gas
284	SiH2Br2	DIBROMOSILANE	25.760	1.7618E-01	-1.6179E-04	7.0860E-08	-1.1902E-11	100	1500	gas
285	SiH2Cl2	DICHLOROSILANE	21.583	1.7618E-01	-1.6179E-04	7.0860E-08	-1.1902E-11	100	1500	gas
286	SiH2F2	DIFLUOROSILANE	6.367	1.9374E-01	-1.6805E-04	7.1483E-08	-1.1961E-11	298	1500	gas
287	SiH2I2	DIIODOSILANE	27.010	1.7618E-01	-1.6179E-04	7.0860E-08	-1.1902E-11	100	1500	gas
288	SiH3Br	MONOBROMOSILANE	10.745	1.9428E-01	-1.9487E-04	1.0614E-07	-2.3863E-11	298	1500	gas
289	SiH3Cl	MONOCHLOROSILANE	7.830	1.9428E-01	-1.9487E-04	1.0614E-07	-2.3863E-11	298	1500	gas
290	SiH3F	MONOFLUOROSILANE	3.610	1.9169E-01	-1.7363E-04	8.3890E-08	-1.6955E-11	298	1500	gas
291	SiH3I	IODOSILANE	9.485	1.9428E-01	-1.9487E-04	1.0614E-07	-2.3863E-11	298	1500	gas
292	SiH4	SILANE	28.887	1.7546E-02	1.4919E-04	-1.5680E-07	4.6291E-11	100	1500	gas
293	SiO2	SILICON DIOXIDE	-7.170	2.5900E-01	-3.3460E-04	1.6970E-07	0.0000E+00	100	848	solid
294	Si2Cl6	HEXACHLORODISILANE*	103.559	3.1545E-01	-5.0601E-04	3.5370E-07	-8.9635E-11	100	1500	gas
295	Si2F6	HEXAFLUORODISILANE*	76.975	2.7359E-01	-3.5566E-04	2.1540E-07	-4.9404E-11	100	1500	gas
296	Si2H5Cl	DISILANYL CHLORIDE*	48.881	1.9428E-01	-1.9487E-04	1.0614E-07	-2.3863E-11	298	1500	gas
297	Si2H6	DISILANE	27.353	1.9208E-01	-5.8767E-05	-3.4180E-08	1.8348E-11	100	1500	gas
298	Si2OCl3F3	TRICHLOROTRIFLUORODISILOXANE*	101.566	2.1385E-01	-3.2354E-04	2.2226E-07	-5.6419E-11	298	1500	gas
299	Si2OCl6	HEXACHLORODISILOXANE*	96.365	3.1545E-01	-5.0601E-04	3.5370E-07	-8.9635E-11	100	1500	gas
300	Si2OH6	DISILOXANE*	37.295	1.9208E-01	-5.8767E-05	-3.4180E-08	1.8348E-11	100	1500	gas
301	Si3Cl8	OCTACHLOROTRISILANE*	141.065	3.1545E-01	-5.0601E-04	3.5370E-07	-8.9635E-11	100	1500	gas
302	Si3H8	TRISILANE*	47.295	1.9208E-01	-5.8767E-05	-3.4180E-08	1.8348E-11	100	1500	gas
303	Si3H9N	TRISILAZANE*	81.085	1.9208E-01	-5.8767E-05	-3.4180E-08	1.8348E-11	100	1500	gas
304	Si4H10	TETRASILANE*	93.015	1.9208E-01	-5.8767E-05	-3.4180E-08	1.8348E-11	100	1500	gas
305	Sm	SAMARIUM	13.224	7.5137E-02	-8.6053E-05	6.5667E-08	-1.8682E-11	298	1190	solid
306	Sn	TIN	21.589	1.8159E-02	-3.1578E-12	5.2788E-15	-3.2746E-18	298	505	solid
307	SnBr4	STANNIC BROMIDE*	46.415	3.1545E-01	-5.0601E-04	3.5370E-07	-8.9635E-11	100	1500	gas
308	SnCl2	STANNOUS CHLORIDE	67.781	3.8744E-02	4.4328E-13	-7.1475E-16	4.2727E-19	298	520	solid
309	SnCl4	STANNIC CHLORIDE*	38.672	3.1545E-01	-5.0601E-04	3.5370E-07	-8.9635E-11	100	1500	gas
310	SnH4	STANNIC HYDRIDE*	37.102	1.7546E-02	1.4919E-04	-1.5680E-07	4.6291E-11	100	1500	gas
311	SnI4	STANNIC IODIDE	81.170	1.5062E-01	-2.7687E-13	2.4060E-16	2.3117E-20	298	418	solid
312	Sr	STRONTIUM	22.217	1.3891E-02	2.4603E-16	-3.2509E-19	1.5174E-22	298	862	solid
313	SrO	STRONTIUM OXIDE	18.334	1.4155E-01	-2.2247E-04	1.6187E-07	-4.3754E-11	298	1270	solid
314	Ta	TANTALUM	25.020	2.4853E-03	1.7796E-17	-7.2588E-21	9.9371E-25	298	3269	solid
315	Tc	TECNNETIUM	21.757	8.3680E-03	4.2956E-18	-2.0384E-21	3.3152E-25	298	2473	solid
316	Te	TELLURIUM	19.163	2.1966E-02	2.1203E-14	-2.9036E-17	1.4440E-20	273	723	solid
317	TeCl4	TELLURIUM TETRACHLORIDE	138.909	-7.9020E-09	3.0372E-11	-5.1354E-14	3.2239E-17	298	497	solid
318	TeF6	TELLURIUM HEXAFLUORIDE*	-5.933	5.1224E-01	-6.4878E-04	3.7509E-07	-8.1524E-11	100	1500	gas
319	Ti	TITANIUM	22.158	1.0284E-02	2.1941E-15	-2.1123E-18	7.2113E-22	298	1155	solid
320	TiCl4	TITANIUM TETRACHLORIDE	67.914	6.6801E-02	-3.5178E-05	7.2884E-09	-5.1766E-13	100	6000	gas
321	Tl	THALLIUM	43.704	-1.6590E-01	5.4479E-04	-7.2978E-07	3.7871E-10	298	507	solid
322	TlBr	THALLOUS BROMIDE	41.631	2.9706E-02	4.6309E-14	-6.1964E-17	3.0195E-20	298	733	solid
323	TlI	THALLOUS IODIDE	48.367	1.3891E-02	5.1273E-11	-9.3369E-14	6.3394E-17	298	438	solid
324	Tm	THULIUM	25.104	6.2760E-03	1.4555E-16	-9.3336E-20	2.0665E-23	298	1900	solid
325	U	URANIUM	38.834	-9.8635E-02	2.6870E-04	-2.4221E-07	8.2396E-11	298	941	solid
326	UF6	URANIUM FLUORIDE*	-4.133	5.1224E-01	-6.4878E-04	3.7509E-07	-8.1524E-11	100	1500	gas

** A computer program, containing data for all compounds, is available for a nominal fee (Carl L. Yaws, Box 10053, Lamar University, Beaumont, TX 77710, phone/FAX 409-880-8787). The computer program is in ASCII which can be accessed by other software.

```
----------------------------------------------------------------------------------------------------
                                      Cp = A + B T + C T² + D T³ + E T⁴    (Cp - joule/g-mol K, T - K)
                                      ----------------------------------------------------------------
NO   FORMULA   NAME                   A        B          C          D           E         TMIN TMAX PHASE
---- --------- ----------             ------   --------   --------   --------    --------   ---- ---- -----
327  V         VANADIUM               22.898   3.6997E-03  7.8595E-06 -3.7455E-09  6.4469E-13  298 2190 solid
328  VCl4      VANADIUM TETRACHLORIDE  35.481   3.4358E-01 -6.0413E-04  4.4984E-07 -1.1924E-10   50 1500 gas
329  VOCl3     VANADIUM OXYTRICHLORIDE 29.050   3.2969E-01 -5.2847E-04  3.6323E-07 -8.9429E-11   50 1600 gas
330  W         TUNGSTEN               22.912   4.6861E-03  1.4921E-17 -7.3117E-21  1.2276E-24  298 2500 solid
331  WF6       TUNGSTEN FLUORIDE      35.463   4.3695E-01 -6.4406E-04  4.3271E-07 -1.0794E-10  298 1500 gas
332  Xe        XENON                  20.786   0.0000E+00  0.0000E+00  0.0000E+00  0.0000E+00   50 1500 gas
333  Yb        YTTERBIUM              22.635   8.2843E-03  6.3018E-16 -6.4801E-19  2.3706E-22  298 1071 solid
334  Yt        YTTRIUM                23.932   4.1840E-03 -9.4660E-17  6.6643E-20 -1.6156E-23  298 1773 solid
335  Zn        ZINC                   22.384   1.0042E-02 -5.3572E-14  7.3244E-17 -3.6627E-20  298  693 solid
336  ZnCl2     ZINC CHLORIDE          60.668   2.3012E-02  7.6380E-13 -1.1605E-15  6.4990E-19  298  591 solid
337  ZnF2      ZINC FLUORIDE          62.300   1.1360E-02 -5.2291E-15  5.0388E-18 -1.7239E-21  298 1145 solid
338  ZnO       ZINC OXIDE             -5.070   2.8610E-01 -5.7750E-04  4.3300E-07  0.0000E+00   40  473 solid
339  ZnSO4     ZINC SULFATE           71.427   8.7015E-02  0.0000E+00  0.0000E+00  0.0000E+00  300  953 solid
340  Zr        ZIRCONIUM              21.974   1.1632E-02 -2.1446E-15  2.0750E-18 -7.1349E-22  298 1135 solid
341  ZrBr4     ZIRCONIUM BROMIDE      70.795   3.5625E-01 -8.1803E-04  8.8208E-07 -3.6360E-10  298  720 solid
342  ZrCl4     ZIRCONIUM CHLORIDE     44.472   5.0671E-01 -1.1887E-03  1.2934E-06 -5.3834E-10  298  710 solid
343  ZrI4      ZIRCONIUM IODIDE       92.902   2.0270E-01 -4.6822E-04  5.1226E-07 -2.1446E-10  298  704 solid
----------------------------------------------------------------------------------------------------
```

** A computer program, containing data for all compounds, is available for a nominal fee (Carl L. Yaws, Box 10053, Lamar University, Beaumont, TX 77710, phone/FAX 409-880-8787). The computer program is in ASCII which can be accessed by other software.

NOTE:

1. Sources for the property data are:

1. Daubert, T. E. and R. P. Danner, DATA COMPILATION OF PROPERTIES OF PURE COMPOUNDS, Parts 1, 2, 3 and 4, Supplements 1 and 2, DIPPR Project, AIChE, New York, NY (1985-1992).
2. Stull, D. R. and H. Prophet, JANAF THERMOCHEMICAL TABLES, 2nd edition, NSRDS-NBS 37, US Government Printing Office, Washington, DC (June, 1971).
3. Wagman, D. D. and others, NBS TABLES OF CHEMICAL THERMODYNAMIC PROPERTIES, J. Phys. Chem. Ref. Data, 11, Supplement No. 2 (1982).
4. Chase, M. W. and others, JANAF THERMOCHEMICAL TABLES, 3rd edition, Parts 1 (Al-Co) and 2 (Cr-Zr), J. Phys. Chem. Ref. Data, 14, Supplement No. 1 (1985).
5. Kelley, K. K., CONTRIBUTIONS TO THE DATA ON THEORETICAL METALLURGY, Bureau of Mines Bulletin 584, US Government Printing Office, Washington, DC (1960).
6. Wicks, C. E. and F. E. Block, THERMODYNAMIC PROPERTIES OF 65 ELEMENTS - THEIR OXIDES, HALIDES, CARBIDES, AND NITRIDES, Bureau of Mines Bulletin 605, US Government Printing Office, Washington, DC (1963).
7. Barin, I. and O. Knacke, THERMOCHEMICAL PROPERTIES OF INORGANIC SUBSTANCES, Springer-Verlag, New York, NY (1973).
8. PERRY'S CHEMICAL ENGINEERING HANDBOOK, 6th ed., McGraw-Hill, New York, NY (1984).
9. CRC HANDBOOK OF CHEMISTRY AND PHYSICS, 66th - 75th eds., CRC Press, Inc., Boca Raton, FL (1985-1994).
10. CONDENSED CHEMICAL DICTIONARY, 10th (G. G. Hawley) and 11th eds. (N. I. Sax and R. J. Lewis, Jr.), Van Nostrand Reinhold Co., New York, NY (1981,1987).
11. LANGE'S HANDBOOK OF CHEMISTRY, 13th and 14th eds., McGraw-Hill, New York, NY (1985, 1992).
12. Reid, R. C., J. M. Prausnitz and B. E. Poling, THE PROPERTIES OF GASES AND LIQUIDS, 3rd ed. (R. C. Reid and T. K. Sherwood), 4th ed., McGraw-Hill, New York, NY (1977, 1987).
13. Rabinovich, V. A., editor, THERMOPHYSICAL PROPERTIES OF GASES AND LIQUIDS, translated from Russian, U. S. Dept. Commerce, Springfield, VA (1970).
14. Yaws, C. L. and others, Solid State Technology, 16, No. 1, 39 (1973).
15. Yaws, C. L. and others, Solid State Technology, 17, No. 1, 47 (1974).
16. Yaws, C. L. and others, Solid State Technology, 17, No. 11, 31 (1974).
17. Yaws, C. L. and others, Solid State Technology, 18, No. 1, 35 (1975).
18. Yaws, C. L. and others, Solid State Technology, 21, No. 1, 43 (1978).
19. Yaws, C. L. and others, Solid State Technology, 24, No. 1, 87 (1981).
20. Yaws, C. L. and others, J. Ch. I. Ch. E., 12, 33 (1981).
21. Yaws, C. L. and others, J. Ch. I. Ch. E., 14, 205 (1983).
22. Yaws, C. L. and others, Ind. Eng. Chem. Process Des. Dev., 23, 48 (1984).
23. Yaws, C. L., PHYSICAL PROPERTIES, McGraw-Hill, New York, NY (1977).
24. Yaws, C. L., THERMODYNAMIC AND PHYSICAL PROPERTY DATA, Gulf Publishing Co., Houston, TX (1992).
25. Yaws, C. L. and R. W. Gallant, PHYSICAL PROPERTIES OF HYDROCARBONS, Vols. 1 (2nd ed.), 2 (3rd ed.), 3, and 4, Gulf Publishing Co., Houston, TX (1992, 1993, 1993, 1995).
26. Estimated.

2. Very limited experimental data for heat capacity are available for inorganic compounds as compared to the more abundant experimental data which are available for organic compounds. Thus, the estimates for these substances should be considered rough approximations in the absence of experimental data. The estimates are noted by the * following the compound name.

Appendix F

COMPOUND LIST BY FORMULA

Appendix G

COMPOUND LIST BY NAME

Appendix H

Computer Program for Thermodynamic Properties

A computer program for calculation of thermodynamic properties using the Peng-Robinson equation of state is available for a nominal fee (Carl L. Yaws, Box 10053, Lamar University, Beaumont, TX 77710, phone/FAX 409-880-8787). The computer program is executable and complete with data files. The program calculates thermodynamic properties at pressures and temperatures that are input by the user. Representative results are shown below:

COMPOUND: 1534 N2   NITROGEN

reference state: datum of ideal gas @ 77 F (25 C)

P psia	T F	Z	V ft^3/lb	H BTU/lb	S BTU/lb F
500.0	-300.00	0.152	0.019	-174.07	-0.892
500.0	0.00	0.970	0.342	-24.31	-0.297
500.0	100.00	0.993	0.426	2.22	-0.245
500.0	200.00	1.003	0.507	28.21	-0.202
500.0	300.00	1.008	0.587	53.98	-0.166
500.0	400.00	1.010	0.666	79.70	-0.134
500.0	500.00	1.012	0.744	105.49	-0.106
500.0	1000.00	1.011	1.131	237.14	0.004
3000.0	-300.00	0.855	0.017	-169.89	-0.917
3000.0	0.00	0.988	0.058	-41.19	-0.456
3000.0	100.00	1.040	0.074	-8.86	-0.392
3000.0	200.00	1.065	0.090	20.95	-0.343
3000.0	300.00	1.076	0.104	49.45	-0.303
3000.0	400.00	1.080	0.119	77.22	-0.268
3000.0	500.00	1.081	0.132	104.59	-0.238
3000.0	1000.00	1.068	0.199	240.67	-0.124
10000.0	-300.00	2.636	0.016	-154.48	-0.956
10000.0	0.00	1.507	0.027	-43.69	-0.562
10000.0	100.00	1.444	0.031	-10.25	-0.496
10000.0	200.00	1.402	0.035	21.55	-0.444
10000.0	300.00	1.369	0.040	52.13	-0.401
10000.0	400.00	1.342	0.044	81.84	-0.364
10000.0	500.00	1.318	0.048	110.95	-0.332
10000.0	1000.00	1.234	0.069	253.15	-0.212

Printed and bound by CPI Group (UK) Ltd, Croydon, CR0 4YY

03/10/2024

01040335-0020